Plant Tissue Culture and Secondary Metabolites Production

Plant Tissue Culture and Secondary Metabolites Production

Editors

Laura Pistelli
Kalina Danova

MDPI • Basel • Beijing • Wuhan • Barcelona • Belgrade • Manchester • Tokyo • Cluj • Tianjin

Editors

Laura Pistelli
University of Pisa
Italy

Kalina Danova
Bulgarian Academy of Sciences
Bulgaria

Editorial Office
MDPI
St. Alban-Anlage 66
4052 Basel, Switzerland

This is a reprint of articles from the Special Issue published online in the open access journal *Plants* (ISSN 2223-7747) (available at: https://www.mdpi.com/journal/plants/special_issues/tissue_metabolites).

For citation purposes, cite each article independently as indicated on the article page online and as indicated below:

LastName, A.A.; LastName, B.B.; LastName, C.C. Article Title. *Journal Name* **Year**, *Volume Number*, Page Range.

ISBN 978-3-0365-6786-0 (Hbk)
ISBN 978-3-0365-6787-7 (PDF)

Cover image courtesy of Kalina Danova

Contents

About the Editors

Laura Pistelli

Laura Pistelli is an assistant professor in plant biology at the University of PISA. Her research deals with the production of specific natural metabolites (antioxidants, nutraceuticals, pharmaceuticals) of aromatic and medicinal plants. The metabolites are isolated from various organs (leaves, roots, flowers, fruits, seeds) and quantified with modern technologies. The tolerance or susceptibility of plants to abiotic stress (e.g., temperature, water availability, pollutants) can promote the production of secondary metabolites. *In vitro* cultures (organ tissues, cell biomass, hairy roots) are also used, which offer a valuable source for optimizing plant proliferation and metabolite production under controlled conditions.

Kalina Danova

Kalina Danova graduated as a Master of Pharmacy from the Faculty of Pharmacy, Medical University, Sofia, Bulgaria. Her PhD on the topic of the in vitro cultivation, physiological behavior, and secondary metabolites production of *Pulsatilla and Hypericum* species and the cryopreservation of *Hypericum rumeliacum* Boiss. was obtained from the Faculty of Plant Physiology, Sofia University, Bulgaria. She is currently an associate professor at the Institute of Organic Chemistry with the Centre of Phytochemistry, Bulgarian Academy of Sciences, and is in charge of the topic of plant cell tissue and organ culture development in medicinal and aromatic plants, with a focus on secondary metabolites and physiological studies as well as germplasm conservation.

plants

MDPI

Editorial

Plant Tissue Culture and Secondary Metabolites Production

Kalina Danova [1,*] and Laura Pistelli [2,*]

1 Institute of Organic Chemistry with Centre of Phytochemistry, Bulgarian Academy of Sciences, Acad. Georgi Bonchev Str., bl.9, 1113 Sofia, Bulgaria
2 Department of Agriculture, Food and Environment, University of Pisa, Via del Borghetto 80, 56124 Pisa, Italy
* Correspondence: k_danova@abv.bg (K.D.); laura.pistelli@unipi.it (L.P.)

Citation: Danova, K.; Pistelli, L. Plant Tissue Culture and Secondary Metabolites Production. *Plants* **2022**, *11*, 3312. https://doi.org/10.3390/plants11233312

Received: 2 November 2022
Accepted: 27 November 2022
Published: 30 November 2022

Plants have developed a complex biochemical system for interacting and coping with dynamic environmental challenges throughout their whole life. Plant secondary metabolites are specifically produced and accumulated in low quantities in response to numerous factors such as bacteria, viruses, fungi, nematodes, insects and herbivores, as well as to climatic factors and seasonal fluctuations, soil and water parameters, etc.

To optimize secondary metabolites production, plants have fine-tuned early signal systems for differentiating general mechanical damage from an attack by an insect/herbivore, for example. In addition, plants distinguish the degree of damage caused by herbivore feeding guilds, insect oral secretions, oviposition fluids, etc. The consecutive steps of the production of the respective defense secondary metabolites is mediated by cellular messengers and events such as metabolic changes, gene activation, jasmonic acid (JA) accumulation, kinase cascades, hydrogen peroxide production, cytosolic calcium ion fluxes, as well as membrane potential changes [1] and references cited therein.

The key role of secondary metabolites for plant survival also underline the pharmacological roles that these substances play in mammalian organisms and hence their applicability in veterinarian and humanitarian medicinal practices.

Plant cell tissue and organ cultures are based on the "totipotency" of the plant cell and its capability to regenerate up to a whole integral organism. The technique allows for the cultivation of separate cells, tissues, differentiated organs, or integral plants in a growth medium in sterile conditions and out of the indigenous natural environment of the plants. The contemporary development of the method has nowadays led to its routine use as a supplementary to conventional plant breeding for an array of applications such as the rapid and disease-free micropropagation of plantlets, independent of seasonal, climatic, and geographic factors; the rapid testing and practical introduction of new cultivars using conventional or genetic engineering selection approaches. An important and rapidly developing field of the application of plant biotechnology is its use for the yield of plant secondary metabolites, allowing for the standardization of the levels of the secondary metabolites produced due to the capability of tissue culture techniques to optimize culture conditions and obtain the desired environment for the production of the target compounds.

Considering the importance of plant secondary metabolites and the high relevance of the establishment of scientifically based approaches for their biotechnological production, we are pleased to present this Special Issue of *Plants* dedicated to "Plant Tissue Culture and Secondary Metabolites Production". The present collection aims to provide readers with up-to-date research dedicated to the scientific accomplishments in the production of plant secondary metabolites of different chemical types through the development of plant cells, tissues, and organs in diverse in vitro culture systems.

We have received seven scientific research papers on tissue culture development and secondary metabolite production of medicinal and aromatic plants of different regions of the world.

Ramabulana et al. [2] established that the exogenous application of auxin (2,4-dichlorop henoxyacetic acid—2,4-D) and cytokinin (benzylaminopurine, BAP) could induce the manipulation of the metabolome of the callus culture of *Bidens pilosa* L. (Asteraceae), dominated by chlorogenic acids consisting of caffeoyl and feruloyl derivatives of quinic acid. Xanthones production in differentiated shoot cultures of the endangered *Gentianella lutescens* (Gentianeceae) was evaluated for the first time by Krstić-Milošević et al. [3] in an experiment focusing on the modification of the concentration of sucrose, sorbitol, and abiotic elicitors salicylic acid (SA), jasmonic acid (JA), and methyl jasmonate (MeJA). Wojtania and Mieszczakowska-Frąc [4] proposed an efficient biotechnological method to produce anthocyanins-rich planting material for selected genotypes of the Polish Culinary rhubarb 'Malinowy' cultivar. In a carrot (*Daucus carota*, Apiaceae) callus culture experimental design of a combination of the components of the Gamborg [5] and Murashige and Skoog [6] culture media, Oleszkiewicz et al. [7] established that the N concentration and the NO_3:NH_4 ratio affected carotenoid accumulation. In this work, changes to the medium other than N, such as microelements, vitamins, growth regulators, and sucrose, had no effect on callus growth and carotenoid accumulation. Mamdouh et al. [8] developed an in vitro protocol for micropropagation of *Lycium schweinfurthii* (Solanaceae). Investigations on genetic stability, phenolic, flavonoid, ferulic acid contents, and antioxidant activity were performed, leading to the selection of an effective protocol for the in vitro propagation of plant material with desired quality. Erst et al. [9] studied the combined effects of NO_3, as well as NH_4:K^+ ratio and the cytokinins BAP and naphthylacetic acid (NAA) on the growth and production of total phenolics callus culture of *Rhodiola rosea* (Crassulaceae). The study of Pieracci et al. [10] on shoot cultures of halophyte *Artemisia caerulescens* L. (Asteraceae) demonstrated the potential of the tissue culture technique for both the ex situ conservation and production of essential oils and phenolic compounds of this species.

Author Contributions: All authors contributed equally to this article. All authors have read and agreed to the published version of the manuscript.

Funding: This research received no external funding.

Data Availability Statement: Not appliable.

Conflicts of Interest: The authors declare no conflict of interest.

References

1. Fürstenberg-Hägg, J.; Zagrobelny, M.; Bak, S. Plant Defense against Insect Herbivores. *Int. J. Mol. Sci.* **2013**, *14*, 10242–10297. [CrossRef] [PubMed]
2. Ramabulana, A.-T.; Steenkamp, P.A.; Madala, N.E.; Dubery, I.A. Application of Plant Growth Regulators Modulates the Profile of Chlorogenic Acids in Cultured *Bidens pilosa* Cells. *Plants* **2021**, *10*, 437. [CrossRef] [PubMed]
3. Krstić-Milošević, D.; Banjac, N.; Janković, T.; Vinterhalter, D.; Vinterhalter, B. *Gentianella lutescens* subsp. carpatica J. Holub.: Shoot Propagation In Vitro and Effect of Sucrose and Elicitors on Xanthones Production. *Plants* **2021**, *10*, 1651. [CrossRef] [PubMed]
4. Wojtania, A.; Mieszczakowska-Frąc, M. In Vitro Propagation Method for Production of Phenolic-Rich Planting Material of Culinary Rhubarb 'Malinowy'. *Plants* **2021**, *10*, 1768. [CrossRef] [PubMed]
5. Gamborg, O.L.; Miller, R.A.; Ojima, K. Nutrient requirements of suspension cultures of soybean root cells. *Exp. Cell Res.* **1968**, *50*, 151–158. [CrossRef] [PubMed]
6. Murashige, T.; Skoog, F. A Revised Medium for Rapid Growth and Bioassays with Tobacco Tissue Cultures. *Physiol. Plant.* **1962**, *15*, 473–497. [CrossRef]
7. Oleszkiewicz, T.; Kruczek, M.; Baranski, R. Repression of Carotenoid Accumulation by Nitrogen and NH_4^+ Supply in Carrot Callus Cells In Vitro. *Plants* **2021**, *10*, 1813. [CrossRef] [PubMed]
8. Mamdouh, D.; Mahgoub, H.A.M.; Gabr, A.M.M.; Ewais, E.A.; Smetanska, I. Genetic Stability, Phenolic, Flavonoid, Ferulic Acid Contents, and Antioxidant Activity of Micropropagated *Lycium schweinfurthii* Plants. *Plants* **2021**, *10*, 2089. [CrossRef] [PubMed]
9. Erst, A.A.; Petruk, A.A.; Erst, A.S.; Krivenko, D.A.; Filinova, N.V.; Maltseva, S.Y.; Kulikovskiy, M.S.; Banaev, E.V. Optimization of Biomass Accumulation and Production of Phenolic Compounds in Callus Cultures of *Rhodiola rosea* L. Using Design of Experiments. *Plants* **2022**, *11*, 124. [CrossRef] [PubMed]
10. Pieracci, Y.; Vento, M.; Pistelli, L.; Lombardi, T.; Pistelli, L. Halophyte *Artemisia caerulescens* L.: Metabolites from In Vitro Shoots and Wild Plants. *Plants* **2022**, *11*, 1081. [CrossRef] [PubMed]

plants

Article

Application of Plant Growth Regulators Modulates the Profile of Chlorogenic Acids in Cultured *Bidens pilosa* Cells

Anza-Tshilidzi Ramabulana [1], Paul A. Steenkamp [1], Ntakadzeni E. Madala [1,2] and Ian A. Dubery [1,*]

[1] Department of Biochemistry, Research Centre for Plant Metabolomics, University of Johannesburg, Auckland Park, Johannesburg 2006, South Africa; 201404841@student.uj.ac.za (A.-T.R.); psteenkamp@uj.ac.za (P.A.S.); ntaka.madala@univen.ac.za (N.E.M.)

[2] Department of Biochemistry, University of Venda, Thohoyandou 0950, South Africa

* Correspondence: idubery@uj.ac.za; Tel.: +27-11-5592401

Abstract: Plant cell culture offers an alternative to whole plants for the production of biologically important specialised metabolites. In cultured plant cells, manipulation by auxin and cytokinin plant growth regulators (PGRs) may lead to in vitro organogenesis and metabolome changes. In this study, six different combination ratios of 2,4-dichlorophenoxyacetic acid (2,4-D) and benzylaminopurine (BAP) were investigated with the aim to induce indirect organogenesis from *Bidens pilosa* callus and to investigate the associated induced changes in the metabolomes of these calli. Phenotypic appearance of the calli and total phenolic contents of hydromethanolic extracts indicated underlying biochemical differences that were investigated using untargeted metabolomics, based on ultra-high-performance liquid chromatography quadrupole time-of-flight mass spectrometry (UHPLC–qTOF–MS), combined with multivariate data analysis. The concentration and combination ratios of PGRs were shown to induce differential metabolic responses and, thus, distinct metabolomic profiles, dominated by chlorogenic acids consisting of caffeoyl- and feruloyl-derivatives of quinic acid. Although organogenesis was not achieved, the results demonstrate that exogenous application PGRs can be used to manipulate the metabolome of *B. pilosa* for in vitro production of specialised metabolites with purported pharmacological properties.

Keywords: auxin; *Bidens pilosa*; cytokinin; callus; chlorogenic acids; organogenesis; phenolics; secondary metabolites

Citation: Ramabulana, A.-T.; Steenkamp, P.A.; Madala, N.E.; Dubery, I.A. Application of Plant Growth Regulators Modulates the Profile of Chlorogenic Acids in Cultured *Bidens pilosa* Cells. *Plants* **2021**, *10*, 437. https://doi.org/10.3390/plants10030437

Academic Editor: Laura Pistelli

Received: 26 January 2021
Accepted: 11 February 2021
Published: 25 February 2021

1. Introduction

Plant secondary (specialised) metabolites are distinctive sources to pharmaceuticals, food additives, flavours, and medicines [1,2]. *Bidens pilosa* L. is a widely occurring annual species of herbaceous flowering plants in the Asteraceae family and consumed as a leafy green vegetable. The plant is also noted for its medicinal value, containing a wide spectrum of natural products, which include aliphatics, aromatic compounds, terpenoids, flavonoids, hydroxycinnamic acids (HCAs), and HCA derivatives such as chlorogenic acids (CGAs) that may be synthesised as a number of distinct regio- or geometrical isomers [3–5]. In previous studies, CGAs were reported to have various health benefits such as anti-diabetic properties [6], HIV-integrase inhibition by 3,5-*di*-caffeoylquinic acid [7] and anti-cancer properties linked to 4,5-*di*-caffeoylquinic acid [8]. In addition to differentiated stem and leaf tissues, undifferentiated cultured cells of *Bidens pilosa* were also found to produce some of these bioactive specialised metabolites [5].

Although this plant contains a wide variety of important metabolites, the natural habitats of plants are being lost due to agricultural practices, urbanization, and other environmental disturbances such as global warming. Therefore, the use of plant cell culture has gained interest in the sustainable and conservative production of various bioactive plant specialised metabolites [9]. These in vitro systems offer the ability to produce high-value marketable natural products at high yields with consistent quality at shorter production cycles [10], overcoming inconveniences of using whole plants.

In cell culture, plant growth regulators (PGRs) are added to basal growth media to stimulate developmental responses. Of these, auxins and cytokinins are the most frequently used PGRs for most applications [11]. When added exogenously, they interact with other endogenously produced phytohormones to modulate developmental processes [12] such as the formation of meristems [13], that contain a small group of pluripotent stem cells, which are responsible for formation of all tissues of a plant [14,15]. These hormones also act synergistically in controlling cell division in undifferentiated cells [16].

In tissue culture, plant cells can be manipulated to regenerate plant tissues from somatic differentiated cells under favourable conditions (organogenesis). This is achieved over a two-step process, which includes acquisition of pluripotent cells followed by de novo shoot organogenesis [17]. Generally, a low auxin: cytokinin ratio promotes shoot induction (caulogenesis) from callus cells, while high auxin: cytokinin ratio promotes root formation (rhizogenesis) [14,18–20]. In vitro organogenesis can be described as either indirect or direct. In the former, PGRs stimulate totipotent cells of callus for organogenesis [21], while direct organogenesis entails plant regeneration directly from explant material [22]. The mode of crosstalk between these PGRs varies with plant species and organs being studied [13]. The use of PGRs in cell culture may also elicit production of specialised metabolites that assist cells to adapt to and survive in in vitro growth conditions [20]. Consequently, in this study, the effects of different ratios of PGRs were investigated on *B. pilosa* callus derived from stems and leaf tissues in order to evaluate how PGR-mediated manipulation of the undifferentiated callus cells would affect the phytochemical profiles of the CGAs previously recorded in *B. pilosa* tissues [5] and the regenerative potential of these cells.

2. Results and Discussion

2.1. Manipulation of Undifferentiated *Bidens pilosa* Cells with Plant Growth Regulators

Organogenesis allows for control of plant development and production of specific tissues in vitro for specialised metabolite biosynthesis [23]. *B. pilosa* stems and leaves explant materials were used to initiate callus on media with different combinations of 2,4-D (auxin) and benzylaminopurine (BAP) (cytokinin). Good callus formation was observed on the combination of 0.45 mg/L 2,4-D and 1.0 mg/L BAP. Calli that formed under these conditions were cut from the original explants and transferred to fresh medium of the same composition. Stem- and leaf-derived calli were cultured until cell growth stabilised post multiple subsequent sub-culturing steps. The white friable callus was sub-cultured onto solid medium with different combinations of auxins and cytokinins (2,4-D: BAP) as detailed in Table 1 to stimulate the undifferentiated cells towards root—or shoot—organogenesis.

Table 1. Concentration ratios of auxin (2,4-D, dichlorophenoxyacetic acid) and cytokinin (BAP, benzylamino purine) used to cultivate *Bidens pilosa* callus on Murashige and Skoog (MS) media.

Condition Number	2,4-D (mg/L)	BAP (mg/L)	Ratio (2,4-D: BAP)
1	0.20	2.00	1:10
2	2.00	0.20	10:1
3	0.00	0.00	-
4	0.45	1.00	1:2
5	0.30	4.00	1:20
6	0.20	8.00	1:40

B. pilosa leaf callus was found to grow well into friable white callus (Figure 1(A1)) in response to a combination of low auxin and high cytokinin concentrations (0.2 mg/L 2,4-D and 2.0 mg/L BAP), respectively. Similarly, leaf callus grown on PGR ratios with very high cytokinin concentrations (0.3 mg/L 2,4-D: 4.0 mg/L BAP) (Figure 1(A5)) and (0.2 mg/L 2,4-D: 8.0 mg/L BAP) (Figure 1(A6)) still maintained growth with minor browning. Furthermore, stem callus maintained in media with low auxin and high cytokinin also grew well as seen with leaf callus (Figure 1(B1,B5,B6)). In contrast, leaf callus maintained at high auxin and low cytokinin concentration (2.0 mg/L 2,4-D: 0.2 mg/L BAP) (Figure 1(A2)) showed comparably reduced growth and higher levels of callus browning, whilst stem callus grew considerably better under the same conditions (Figure 1(B2)). Similarly, callus browning and reduced growth (Figure 1A3) was also observed in callus maintained in the initiation media (0.45 mg/L 2,4-D:1.0 mg/L BAP), contrasting with stem callus that maintained growth and was not oxidised (Figure 1(B3)). Callus browning/oxidation has been attributed to many factors that are correlated to an increase in phenolic content. The increase in phenolic content is related to an increase in the activity of the enzyme phenylalanine ammonia-lyase (PAL), which converts phenylalanine to trans-cinnamic acid, subsequently leading to biosynthesis of cinnamate derivatives, which have been associated with browning [24]. This coincides with elevated activity of enzymes such as polyphenol oxidase [25] and peroxidase [26], which convert phenolics to molecules that have a deleterious effect on callus cultures. Browning has also been correlated to carbohydrate metabolism in cell cultures. Callus browning has been described as a major problem that inhibits shoot formation and long-term maintenance of callus as observed in this study [27].

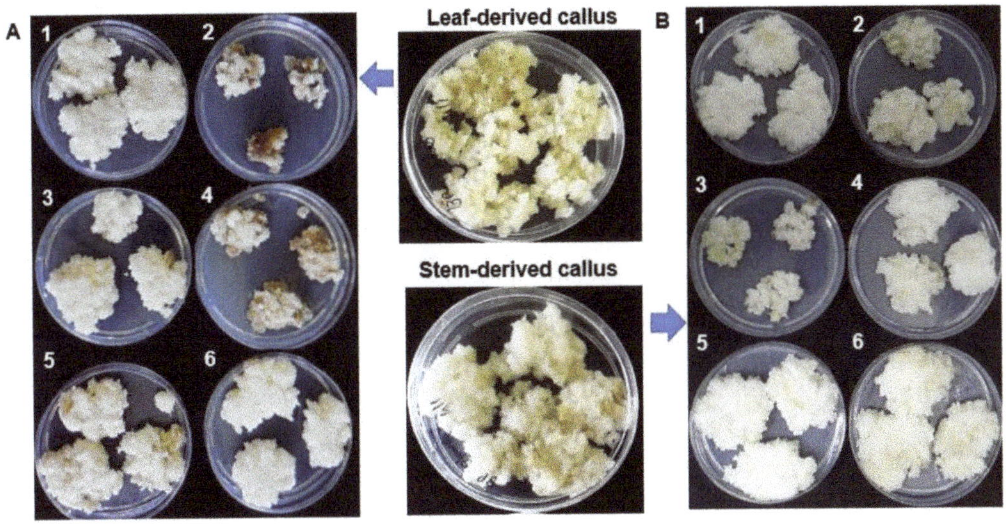

Figure 1. Representative images of *Bidens pilosa* friable stem- and leaf callus initiated on 0.45 mg/L 2,4-dichlorophenoxyacetic acid (2,4-D) and 1.0 mg/L benzyl aminopurine (BAP), which was sub-cultured onto media with different combinations of concentrations/ratios of 2,4-D and BAP. (**A**) Indicates leaf-derived callus maintained on media containing 2,4-D and BAP at the following ratios (1:10) (1), (10:1) (2), (0:0) (3), (1:2) (4), (1:20) (5) and (1:40) (6). (**B**) Indicates stem-derived callus also maintained on media with different 2,4-D: BAP ratios as described for the leaf callus in (A). The details of the plant growth regulator (PGR) concentrations are described in Table 1. The calli were harvested at 25 d following sub-culture, when the first signs of browning started to appear.

Partial callus habituation was observed (Figure 1(A3)), where leaf callus grown on MS medium with organics, without added PGRs, still maintained growth. Habituation is an occurrence in which division and growth of cells in culture become independent of added PGRs [20,28]. This was, however, not observed on stem callus maintained on PGR-free media, where reduced callus growth was observed (Figure 1(B3)).

In this study, *B. pilosa* cell culture did not show signs of possible shoot and root regeneration under the conditions investigated. The different media conditions resulted in either callus browning or white friable cells. Various factors such as multiple sub-culturing are known to result in loss of the totipotency of tissue cultured. Even where explant tissue originated from tissue with identical genetic backgrounds, epigenetic changes may still occur in cultured cells where genes responsible for organogenesis are modified (hypo- and hypermethylated) [29]. Other factors contributing to loss of totipotency and loss of regenerative abilities include genetic variation and selection pressure, resulting in loss of cells with genetic memory for totipotency [20,29]. The age of plants used also affects the regenerative capacity of ex-plants, such that explants from younger plants have been found to possess better regenerative capacity compared to matured plants [30]. In future, de novo organogenesis of *B. pilosa* could be studied at different developmental stages of the plant. Culture conditions also play a role in regeneration of the plant, such as media nutrient content, pH, light and temperature. In consideration, further optimisation of the auxin and cytokinin PGRs and maintenance conditions of *B. pilosa* callus can be investigated to achieve shoot and root regeneration if desired [31].

2.2. Total Phenolic Content in Response to Different Plant Growth Regulator Combinations

Phenolic compounds are produced by plants in response to a number of ecological pressures such as biotic and abiotic stresses, but also physiological stress [32]. As plant protective metabolites, phenolic compounds exhibit redox properties, allowing their action as antioxidants. The total phenolic compound (TPC) content assay, thus, reflects the reducing capacity of the extracts and is also used as a general indicator of specialised metabolite synthesis in plants [33]. As reported in Table 2, TPC varied from 18–33 mg gallic acid equivalents (GAE)/g tissue with the lowest determined for leaf-derived calli (condition 1) and the highest for stem-derived calli (condition 5). This could indicate that to achieve high phenolics content in stem calli of *B. pilosa*, moderate high cytokinin to low auxin ratio (condition 5) is required. However, very high cytokinin concentration (condition 6) reduced the phenolics content. Interestingly in leaf calli, the highest TPC was achieved independently of PGRs. In general, the TPC of the stem-derived calli was higher than that of the leaf-derived calli for each condition, and no correlation between browning of the calli and TPC could be observed.

Table 2. Total phenolics content (TPC, expressed as mg gallic acid equivalents/g wet weight) in leaf and stem callus maintained on different concentration ratios of auxin (2,4-dichlorophenoxy-acetic acid, 2,4-D) to cytokinin (benzyl aminopurine, BAP).

Condition Number	2,4-D (mg/L)	BAP (mg/L)	TPC (Leaf Calli)	TPC (Stem Calli)
1	0.20	2.00	18.02 ± 0.08	25.86 ± 0.07
2	2.00	0.20	22.25 ± 0.04	28.29 ± 0.02
3	0.00	0.00	29.20 ± 0.06	31.15 ± 0.02
4 *	0.45	1.00	20.38 ± 0.02	31.15 ± 0.03
5	0.30	4.00	27.12 ± 0.07	33.30 ± 0.05
6	0.20	8.00	25.16 ± 0.08	25.30 ± 0.05

* Condition used for initial establishment of calli. Calli were harvested at 25 d following sub-culture on the new media.

2.3. Analysis of Altered Callus Metabolomes in Response to Different Plant Growth Regulator Combinations

The CGA composition of methanol extracts from tissues, callus and cell suspensions from *B. pilosa* was previously profiled by ultra-high-performance liquid chromatography quadrupole time-of-flight mass spectrometry (UHPLC–qTOF–MS), using an optimized in-source collision-induced dissociation (ISCID) method capable of discriminating between closely related HCA derivatives of quinic acids, based on MS fragmentation patterns [5,34]. This ISCID approach was shown to efficiently discriminate between positional isomers of CGAs. A combined total of 30 *mono-*, *di-*, and *tri*-substituted CGAs were annotated.

The manipulation of PGRs ratios in plant cell cultures does not only affect growth and developmental processes, but also regulates different pathways of specialised metabolism such as the phenylpropanoid pathway [35]. In plant cell culture, a variety of specialised metabolites accumulates differentially depending on the concentrations of PGRs [36]. The PGRs (2,4-D and BAP) caused visible differences in the growth rates of the calli on growth medium that contained the PGRs in different ratios, accompanied with browning of the calli grown under some conditions. These phenotypic differences were also accompanied by differences in the total phenolic content of the tissues, indicative of elicited changes in the metabolomes of the calli due to the auxin: cytokinin PGR ratios. To gain further insights into the underlying changes to the metabolomes, specifically with regards to the CGAs, a targeted metabolomics investigation was performed. The following six different conditions of 2,4-D to BAP were investigated, condition 1 (1:10), condition 2 (10:1), condition 3 (0:0) condition 4 (1:2), condition 5 (1:20) and condition 6 (1:40). The actual concentrations involved are reported in Table 1.

Changes in the metabolome of *B. pilosa* callus were studied with a high-throughput analytical method: UHPLC–qTOF–MS. Through visual inspection of base peak intensity (BPI) chromatograms (Figure 2A,B), HCA derivatives were a notable group of metabolites in leaf and stem callus as observed previously [5,34]. The HCA derivatives were observed to have minor intensity differences between callus grown on media with various ratios of PGRs as shown in yellow rectangles in Figure 2. Although the combined effects of auxins and cytokinins are not fully understood, the addition of PGRs in culture has been shown to generally have positive effects on the accumulation of specialised metabolites, including phenolics [37]. To further visualise the systematic trends in response to the different PGR ratios, the datasets were subjected to multivariate statistical analysis.

2.4. Multivariate Statistical Analysis of Phytochemical Profiles/Constituents of Callus Maintained on Different Plant Growth Regulator Combinations

Unsupervised multivariate statistical analysis was employed to explore datasets generated by the UHPLC–qTOF–MS analysis of methanol extracts of *B. pilosa* stem- and leaf callus maintained on media with different ratios of 2,4-D to BAP ((1:10), (10:1), (0:0), (1:2), (1:20) and (1:40)). To analyse the variability within and between the datasets, PCA scores scatterplot models were constructed for leaf callus (Figure 3A) and stem callus (Figure 3C), which reduced the dimensionality of the datasets [38–40]. The PCA scores scatterplot (Figure 3A) computed for extracts from leaf callus was an 11-component model of which PC1 and PC2 explained 30.6 and 14.5% of the variation within the dataset. Visually, the PCA scores scatterplot indicated differential metabolic profiles within the datasets, as condition-specific clustering was observed. Leaf callus maintained on media with higher cytokinin ((1:10), (1:20) and (1:40) 2,4-D to BAP) were found to be more closely related. The statistical validation of the model was described by the explained variation/goodness of fit $R^2 = 0.876$ and the predictive variance $Q^2 = 0.704$. The statistical validation performed indicated that the model computed was fit, as acceptable models for biological data are described by $R^2 > 0.7$ and $Q^2 > 0.4$ [41].

The PCA scores scatterplot (Figure 3C) computed for stem callus maintained on media with different combination ratios of PGRs indicated differential clustering of sample groups. Visually, stem callus grown on media with (1:2), (1:40) and (10:1) 2,4-D to BAP clustered together and separate from the other conditions. This model was also an 11-component model, where PC1 and PC2 described 30.4 and 19.2% of the variation within the dataset, respectively. The model was found to be adequate to draw relevant biological interpretation described by $R^2 = 0.9$ and the predictive variance $Q^2 = 0.78$. Interestingly, for both stem and leaf callus maintained on medium with no PGRs ((0:0) 2,4-D to BAP), a grouping separate from sample groups grown on media with PGRs was observed. This could indicate that callus grown on media without PGRs (i.e., habituated callus) is significantly different from that grown-on media with PGRs.

Hierarchical cluster analysis (HiCA) was also computed for both stem and leaf callus, which applies an agglomerative ("bottom-up") algorithm to determine correlation/similarities within callus grown on media with different combination ratios of PGRs [42,43]. HiCA dendrograms were generated from the datasets, and these indicated differences in the metabolomic profiles of callus grown on different combination ratios of PGRs for leaf callus (Figure 3B) and stem callus (Figure 3D). The HiCA dendrogram (Figure 3B) indicated that leaf-derived calli maintained on media with high cytokinin to auxin ratios (1:10), (1:20) and (1:40) 2,4-D to BAP were more closely related compared to calli grown on (0:0), (1:2) and (10:1) 2,4-D to BAP media, which clustered together. In stem-derived callus culture, the dendrogram (Figure 3D) indicated that calli maintained on a high cytokinin to auxin ratio ((1:10) and (1:20) 2,4-D to BAP) were more closely related when compared to calli grown on (0:0), (1:2), (1:40) and (10:1) 2,4-D to BAP.

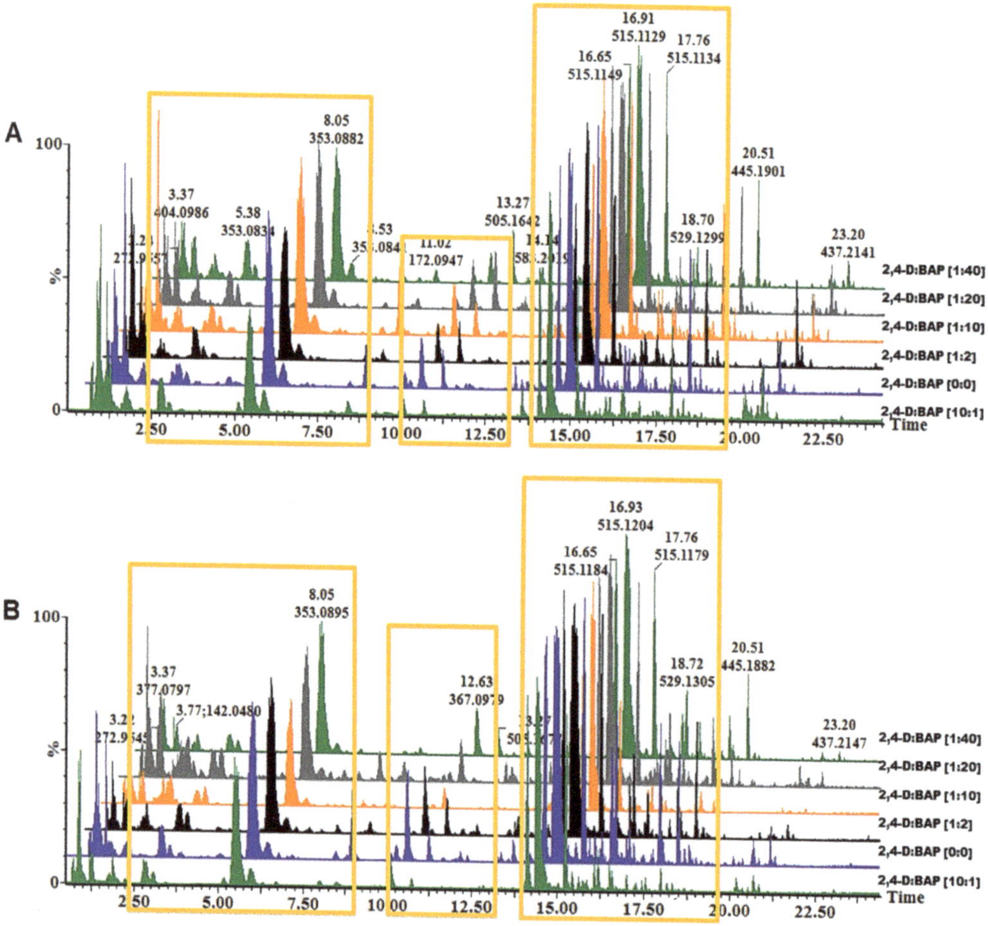

Figure 2. Representative ultra-high-performance liquid chromatography quadrupole time-of-flight mass spectrometry (UHPLC–qTOF–MS) base peak intensity (BPI) chromatograms showing the separation of specialised metabolites in methanol extracts of callus cultures derived from *Bidens pilosa* leaves (**A**) and stems (**B**) maintained on different ratios of 2,4-dichlorophenoxyacetic acid (2,4-D) to benzyl aminopurine (BAP) ((1:10), (10:1), (0:0), (1:2), (1:20) and (1:40)) as detailed in Table 1. Hydroxycinnamic acid derivatives were found to be a prominent group of metabolites in these cultures, albeit with differential peak intensities.

PGRs were also shown to induce differential responses within leaf- and stem-derived callus. In leaf-derived callus, a ratio with very high cytokinin (1:40) 2,4-D to BAP, induced a similar metabolic response as induced by other ratios with high cytokinins ((1:10) and (1:20) 2,4-D to BAP). In contrast, stem-derived callus grown on media with very high cytokinin concentration was shown to have similar metabolic responsiveness as callus grown on media with high auxin concentration ((10:1), 2,4-D to BAP) and moderate cytokinin concentration (1:2) 2,4-D to BAP. This could indicate some tissue-dependent differential responsiveness to the PGRs in the growth media. Further metabolomic differences were assessed post-metabolite identification through relative quantification of the annotated metabolites associated with the different growth conditions.

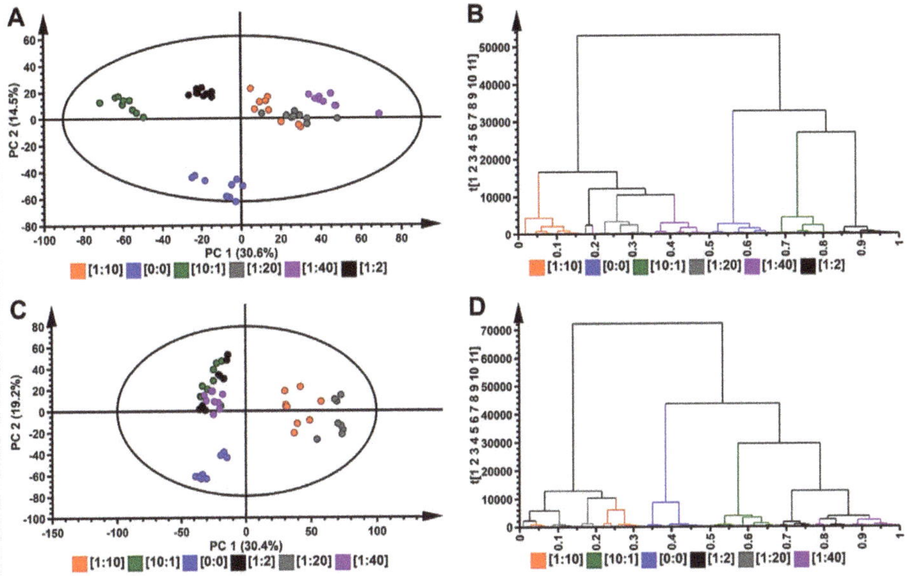

Figure 3. Unsupervised exploratory statistical analysis of *Bidens pilosa* leaf and stem callus maintained on media with different combination ratios of 2,4-dichlorophenoxyacetic acid (2,4-D) to benzyl aminopurine (BAP) ([1:10], [10:1], [0:0], [1:2], [1:20] and [1:40]). (**A**) A principal component analysis (PCA) scores scatterplot of the *Pareto*-scaled dataset of leaf callus. The computed model was an 11-component model, with PC 1 and PC 2 explaining 45.1% of the variation. The quality parameters of the model were: explained variation/goodness of fit R^2 = 0.876 and the predictive variance Q^2 = 0.704. The ellipse in the PCA score scatterplot indicates the Hotelling's T^2 at 95% confidence interval. (**B**) The hierarchical cluster analysis (HiCA) plot shows the hierarchical structure of the data from leaf callus extracts in a dendrogram format, showing plant growth regulator (PGR) concentration/ratio-dependent clustering. (**C**) A PCA score plot scores scatterplot of stem-derived callus. The computed model was an 11-component model, with PC 1 and PC 2 explaining 49.6% of the variation. The quality parameters of the model were: R^2 = 0.9 and Q^2 = 0.78. (**D**) The HiCA plot shows the hierarchical structure of the data from stem-derived callus extracts in a dendrogram format, showing PGR concentration/ratio-dependent clustering.

2.5. Comparative Analysis of Metabolites Identified in Callus Maintained on Media with Different PGR Ratios

Based on the multivariate statistical analysis, metabolites in *B. pilosa* callus grown on solid agar media with different ratios of PGRs, were annotated (putatively identified) to MSI-level 2 as described in [5,34] and listed in Table 3. These were annotated based on accurate mass, retention time (Rt) and mass spectrometric (MS) fragmentation patterns. As previously reported, HCA derivatives such as the CGAs are biologically important metabolites, which have been shown to be abundant in tissues and cell cultures of *B. pilosa* [5]. In this study, metabolites identified from methanol extracts of the callus grown under the mentioned PGRs ratios were found to be primarily HCA derivatives in the form of *mono-* and *di*-acylated quinic acid (i.e., CGAs). A total of 14 CGAs were identified in these callus cultures, consisting of four *mono-*, nine *di-* and one *tri*-substituted quinic acids. These HCAs conjugates to quinic acid did not include coumaric acid and were restricted to caffeic acid and ferulic acid (generating CQA and FQA, respectively). It is also of interest that HCAs conjugated to tartaric acid (caftaric acid and chicoric acid, found in differentiated leaf and stem tissues, [5] were not detected.

Seven other metabolites were found to contribute to the variability between the metabolomic profiles of the various extracts. These were organic acids (gluconic acid, malic acid and citric acid), amino acids (phenylalanine and tryptophan) and two benzoic acids (2,5-dihydroxybenzoic acid and 3,4,5-trihydroxybenzoic acid (as glucogallin)). These metabolites illustrate the link between primary and specialised metabolism with phenylalanine feeding into the phenylpropanoid pathway and being the precursor of the cinnamic and benzoic acids [20,32].

The distribution or accumulation of the HCA derivatives in response to alterations of concentration ratios of 2,4-D to BAP were investigated and highlighted by means of colour-coded heatmaps (Figure 4). These were generated from MetaboAnalyst, in which the resulting heatmaps indicated differential metabolite concentration patterns in response to manipulations with PGRs [44]. A colour gradient was used to indicate abundances of the specialised metabolites, where deep red indicates the highest relative abundances and dark blue indicates the lowest abundances.

Table 3. Characterisation discriminatory metabolites present in methanol extracts of *Bidens pilosa* stem- and leaf-derived callus maintained on media with different combination ratios of 2,4-dichlorophenoxyacetic acid (2,4-D) to benzyl aminopurine (BAP).

No.	*m/z*	Rt (min)	Diagnostic Fragment Ions	Molecular Formulae	Metabolite *	Abbreviation
1	195.590	0.90	191, 162, 108	$C_6H_{12}O_7$	Gluconic acid	Gluc
2	133.010	0.99	115	$C_4H_6O_5$	Malic acid	Mal
3	191.014	1.18	111	$C_6H_8O_7$	Citric acid	CTA
4	331.064	1.72	168, 125	$C_{13}H_{16}O_{10}$	Galloyl-hexoside	Gall
5	164.067	1.96	147	$C_9H_{11}NO_2$	Phenylalanine	Phe
6	315.069	2.07	153, 152, 109, 108	$C_{13}H_{16}O_9$	2,5-Dihydroxybenzoic acid	2,5-DHBA
7	353.0842	2.77	191, 179, 135	$C_{16}H_{18}O_9$	*trans*-3-Caffeoylquinic acid	*trans*-3-CQA
8	203.077	3.22	142, 116	$C_{11}H_{12}N_2O_2$	Tryptophan	Trp
9	353.0881	5.48	191	$C_{16}H_{18}O_9$	*trans*-5-Caffeoylquinic acid	*trans*-5-CQA
10	353.0831	5.87	191, 179, 173, 135	$C_{16}H_{18}O_9$	*trans*-4-Caffeoylquinic acid	*trans*-4-CQA
11	367.0980	10.03	193, 173	$C_{17}H_{20}O_9$	4-Feruloylquinic acid	4-FQA
12	515.1166	14.01	353, 335, 191, 179, 135	$C_{25}H_{24}O_{12}$	3,4-*di*-Caffeoylquinic acid	3,4-*di*CQA
13	515.1195	14.34	353, 191, 179, 135	$C_{25}H_{24}O_{12}$	3,5-*di*-Caffeoylquinic acid	3,5-*di*CQA
14	515.1219	15.17	353, 335, 191, 179, 173, 135	$C_{25}H_{24}O_{12}$	4,5-*di*-Caffeoylquinic acid	4,5-*di*CQA
15	529.1398	15.54	367, 353, 335, 193, 179, 173, 134	$C_{26}H_{26}O_{12}$	3-Feruloyl-4-caffeoylquinic acid	3F-4CQA
16	529.1013	15.72	367, 335, 193, 173	$C_{26}H_{26}O_{12}$	3-Caffeoyl-4-feruloylquinic acid	3C-4FQA
17	529.0983	16.00	367, 193, 134	$C_{26}H_{26}O_{12}$	3-Feruloyl-5-caffeoylquinic acid	3F-5CQA
18	529.1345	16.11	367, 353, 191, 179	$C_{26}H_{26}O_{12}$	3-Caffeoyl-5-feruloylquinic acid	3C-5FQA
19	529.1345	16.53	367, 193, 173	$C_{26}H_{26}O_{12}$	4-Feruloyl-5-caffeoylquinic acid	4F-5CQA
20	529.117	16.68	367, 353, 191, 179, 173, 135	$C_{26}H_{26}O_{12}$	4-Caffeoyl-5-feruloylquinic acid	4C-5FQA
21	677.1561	17.51	515, 353,179, 173	$C_{34}H_{30}O_{15}$	*tri*-Caffeoylquinic acid	*tri*-CQA

* Metabolites with differential *m/z* ion intensities were identified across all conditions in callus derived from both leaves and stems (Figure 4). Callus was harvested after a period of 25 d of growth on media with different concentration ratios of PGRs as listed in Table 1.

In leaf-derived callus (Figure 4A), differential abundance of HCA derivatives was observed in response to alterations of concentration ratios of 2,4-D to BAP ((1:10), (10:1), (0:0), (1:2), (1:20) and (1:40)). Interestingly, leaf callus grown on media without PGRs ((0:0) 2,4-D to BAP) was found to maintain metabolic responsiveness, as for an example 3C-4FQA, 3C-5FQA, 3,4-*di*CQA and *trans*-4-CQA were found to be relatively abundant in this callus type. This could indicate that *B. pilosa* leaf callus could accumulate some HCA derivatives independent of PGRs in culture media. However, other HCA derivatives were also found to be abundant in culture with other combinations of concentration ratios of PGRs ((10:1), (1:2), (1:20) and (1:40) 2,4-D to BAP), which could indicate the requirement of PGRs for accumulation of HCA derivatives in callus culture.

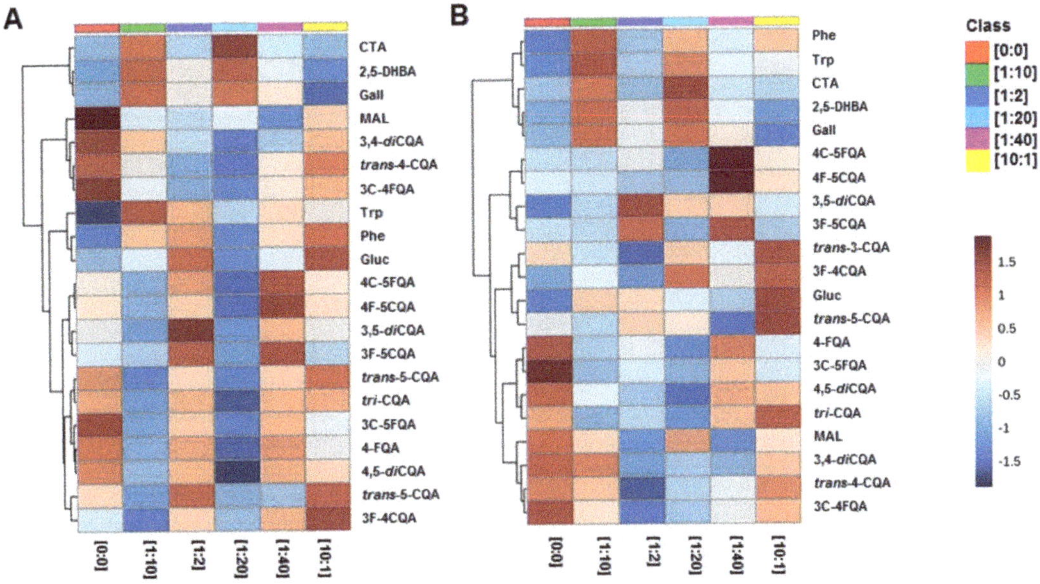

Figure 4. Heatmaps illustrating the occurrence/ distribution of metabolites identified in *B. pilosa* leaf callus (**A**) and stem callus (**B**), maintained on media with different ratios of 2,4-D to BAP ((1:10), (10:1), (0:0), (1:2), (1:20) and (1:40) as detailed in Table 1). Group averages were used to simplify the visualisation of the distribution of these HCA derivatives. Abbreviations of metabolites are as defined in Table 1.

Similarly, stem-derived callus (Figure 4B) abundantly produced some HCA derivatives independent of PGRs ((0:0) 2,4-D to BAP). Generally, some auxins may upregulate production of phenolics [45]. As also observed in this study, high auxin concentration ((10:1) 2,4-D to BAP) resulted in accumulation of *trans*-5-CQA, *trans*-3-CQA and 3F-4CQA in stem-derived callus of *B. pilosa*. Partial suppression of the phenylpropanoid pathway was observed in stem-derived callus grown on media with (1:2) 2,4-D to BAP, as some HCA derivatives were reduced (Figure 4B). This could suggest that other auxin–cytokinin concentration combinations were better for optimal production of HCA derivatives in cell culture of *B. pilosa*. Generally, in both callus culture under conditions with increased HCA derivatives, a decrease in phenylalanine and tyrosine was observed. This could indicate increased activity of phenylalanine/tyrosine ammonia-lyase), a key enzyme in the phenylpropanoid pathway, which is responsible for the conversion of phenylalanine to *trans*-cinnamic acid and tyrosine to *p*-coumaric acid, leading to the subsequent biosynthesis of HCA derivatives [46,47].

3. Materials and Methods

3.1. Callus Initiation and Cultivation on Different Ratios of Plant Growth Regulators

B. pilosa stem and leaf calli were initiated from sterilised explant material taken from plants grown under greenhouse conditions as described previously [5]. Explant material (stem and leaf sections, taken from the same plant) were sterilised with 70% (*v*/*v*) ethanol for 10 s, then 1.5% (*v*/*v*) sodium hypochlorite solution for 20 min and rinsed with sterile distilled water. Cultures were initiated in Petri dishes on Murashige and Skoog (MS) medium [48] supplemented with MS vitamins (0.5 mg/L nicotinic acid, 0.2 mg/L thiamine and 0.5 mg/L pyridoxine). Additions to the MS medium included 100 mg/L myo-inositol, 1 g/L casein hydrolysate, 30 g/L sucrose and 8 g/L phytoagar. The growth medium was supplemented with PGRs: 0.45 mg/L 2,4-dichlorophenoxyacetic acid (2,4-D) and 1 mg/L 6-benzylaminopurine (BAP) at pH 5.8. All PGRs and phytoagar were obtained from Sigma

Aldrich, Muenchen, Germany, while MS medium salts and organics were obtained from Duchefa, (Haarlem, The Netherlands).

Initiated callus was grown in an incubator at 24 °C with a 12 h dark/light cycle and light intensity of 25 µmol/ m^2/ s and sub-cultured onto fresh media every 14 d until callus growth stabilised and the callus was white and friable (approximately 3 months). In order to investigate the effect of combining auxin and cytokinins on undifferentiated *B. pilosa* cells, the friable white calli were sub-cultured onto MS media with vitamins and different concentration ratios of auxin (2,4-D): cytokinin (BAP) as detailed in Table 1. Calli were grown on the new media for 25 d, when signs of browning started to appear on the callus of some of the conditions (Figure 1).

3.2. Metabolite Extraction

Two grams (2 g) of the frozen calli from each condition were weighed and homogenised using a probe Ultra-Turrax homogenizer (IKA, Staufen, Germany) at 100% intensity for 2 min in 20 mL (1:10 *m*/*v*) of 80% analytical grade methanol (Romil SpS Chemistry, Cambridge, UK). A sonicator bath (Branson CPX, Fischer Scientific, Waltham, MA, USA) was used to sonicate the samples at 100% intensity for 30 min prior to centrifugation. The crude extracts were centrifuged in a swinging-bucket benchtop centrifuge (Beckman Coulter, Midrand, South Africa) at 5100 rpm for 20 min. A rotary evaporator (Heidolph Instruments, Schwabach, Germany) was used to evaporate the supernatants under vacuum at 55 °C to approximately 1 mL. Samples were transferred to 2 mL Eppendorf microcentrifuge tubes and dried overnight in a fume hood on a dry bath at 45 °C. The dried residues were then reconstituted with 500 µL of mass spectrometry-grade methanol: milliQ water (1:1, *v*/*v*) in a sonicator bath at 30 °C. The samples were filtered through 0.22 µm nylon filters into glass chromatography vials fitted with 500 µL inserts. For all the treatments, at least three independent biological replicates were prepared. Quality control samples (QCs) were also prepared through pooling together of equal volumes of all the biological replicates. The filtered samples were stored at 4 °C until LC–MS analysis.

3.3. Total Phenolic Content (TPC) Assay

The TPC was determined using the Folin–Ciocalteu (F-C) assay with gallic acid (Sigma Aldrich, Muenchen, Germany) as a calibration standard with concentrations ranging between 250 and 1250 µM [20]. The reconstituted extracts (in 50% methanol), blanks (50% methanol) and a concentration series of the calibration standard dissolved in 50% methanol were each (100 µL) added to 2 mL Eppendorf tubes. To these samples, 200 µL 10% (*v*/*v*) F-C reagent (Sigma Aldrich, Muenchen, Germany) was added, and samples were vortexed thoroughly. A volume of 800 µL of 0.7 M sodium carbonate was then added to the samples and incubated for 2 h at room temperature. A volume of 200 µL of all sample groups in triplicates were transferred to a 96-well microplate, and the absorbance was determined using a microplate reader at 765 nm.

3.4. Ultra-High-Performance Liquid Chromatography—High-Definition Mass Spectrometry (UHPLC–HDMS)

The aqueous-methanol extracts were analysed on an UHPLC high-definition quadrupole time-of-flight MS instrument (UHPLC–qTOF Synapt G1 HDMS system, Waters Corporation, Manchester, UK) for chromatographic separation and mass spectrometric (MS) data acquisition. Prior to MS analysis, the samples were separated on the UHPLC system fitted with an Acquity HSS T3 C18 column (150 × 2.1 mm with particle size of 1.7 µm) (Waters, Milford, MA, USA). A binary solvent system comprising of 0.1% aqueous formic acid in Milli-Q water (solvent A) and 0.1% formic acid in acetonitrile (solvent B) was used. A 30 min binary solvent gradient was run at a flow rate of 0.4 mL/min. The elution gradient conditions were: 2% B over 0.0–1.0 min, 2–60% B over 2.0–24 min, 60–95% B over 24–25 min, from 25–27 min the conditions were held at 95% B, and the column was washed with 95–2% B over 27–28 min. The column was re-equilibrated with 2% B over a 2 min isocratic wash.

The Synapt G1 high definition mass spectrometer, equipped with electrospray ionisation (ESI) source, was used to analyse the separated metabolites by acquiring centroid data in both positive and negative ionisation modes. The conditions for the MS detector were set as: capillary voltage of 2.5 kV, source temperature of 120 °C, sampling cone voltage of 30 V, cone gas flow of 50.0 (L/h), extraction cone of 4.0 V, desolvation gas flow of 550 (L/h), m/z range of 100–1000, scan time of 0.2 s and an interscan delay of 0.02 s. Leucine encephalin $[M + H]^+ = 552.766$ and $[M - H]^- = 554.2615$ was done to ensure high mass was used as a reference calibrant to ensure high mass accuracy (2–5 mDA). The MS analyses were set to result in both unfragmented and fragmented experiments through collision-induced dissociation (MS^E) achieved by alternating the collision energy from 10 to 50 eV. Due to better ionisation in ESI negative mode, only the negative ionisation data were subsequently processed.

3.5. Multivariate Data Analysis, Metabolite Annotation and Relative Quantification

Post data acquisition, negative ionisation MS data were processed using MassLynx XS™ software's MarkerLynx application (Waters, Manchester, UK). Data were processed with the following parameters set: retention time (Rt) range of 0.60–21 min, mass range of 100–200 Da, mass tolerance of 0.05 Da and a Rt window of 0.2 min. Statistical modelling was then performed using the generated data matrix obtained in the "Soft Independent Modelling of Class Analogy" software (SIMCA-15.0, Umetrics Corporation, Umea, Sweden). Pareto scaling was applied on the datasets, whereby data are scaled using the square root of the standard deviation [49]. The statistical models presented here are principal component analysis (PCA) and hierarchical cluster analysis (HiCA), unsupervised models that assess the overall structure of the data and indicate trends, clusters and similarities between sample groups/experimental treatments [50]. Potential biomarkers of different sample groups were highlighted using orthogonal projection to latent structures discriminant analysis (OPLS-DA S-plots), and only significant biomarkers with correlation, $(p(corr)) \geq 0.5$ and covariance, $(p1) \geq 0.05$ were annotated [5,34].

Metabolites were putatively annotated to level 2 of the Metabolomics Standards Initiative (MSI) [51]. The following criteria was utilized: (i) molecular formula from full-scan accurate mass data was used, (ii) the elemental composition predictions were searched against online databases such as Dictionary of Natural Products [52], Kyoto Encyclopedia of Genes and Genomes (KEGG) [53] and Chemspider [54], (iii) fragmentation patterns based on the MS^1 and MS^E spectra of the metabolites were assessed, (iv) putative annotations were also compared to available literature with respect to their reverse-phase column chromatography elution profiles. The relative peak intensities of annotated metabolites were also visualized using colour-coded heatmaps generated from MetaboAnalyst [55]. The data imported were normalised by median, log transformed and Pareto-scaled. The computed HiCA indicated samples with relatively similar abundances. The Pearson's correlation was applied as a dissimilarity measure, and Ward's clustering algorithm was used. To simplify the visualisation of the changing patterns, group averages were used [56].

4. Conclusions

Plant cell cultures of *B. pilosa* were successfully initiated from stem and leaf explant material. Interestingly, metabolomic profiling of such generated callus and cell suspensions demonstrated differential cell-line-specific metabolite distribution, similar to the tissue-specific distribution initially observed in the corresponding differentiated tissues reported previously. This showed that *B. pilosa* cell cultures are a promising alternative approach/source for the production of high-value specialised metabolites such chlorogenic acids, e.g., HCAs esterified to quinic acid (caffeoyl- or feruloyl-quinic acids, CQA and FQA). This study aimed at investigating the effects of auxin (2,4-D) and cytokinin (BAP) on the ability of undifferentiated *B. pilosa* callus cells to regenerate, and to profile the metabolite distribution patterns resulting from the exogenous applications of plant growth regulators. Although the results indicated that treatment with different combinations of auxin and cytokinins did not initiate organogenesis, the callus demonstrated differential accumulation of HCA derivatives in response to the various PGR concentration ratios investigated. The stimulating effects of the PGRs were nuanced in that the results indicated graded differences between the perturbed metabolomes, showing increases for some of the CGAs, but also interconversion between isomeric versions of the same metabolite. This study highlights the differential effects of auxin: cytokinin interactions on the production of specialised metabolites in cultured cells of *B. pilosa* and illustrates the investigation of optimal concentration ratios of PGRs for the biosynthesis of HCA derivatives.

Author Contributions: I.A.D. and N.E.M. conceptualized and supervised the project; A.-T.R. performed the experimental work; P.A.S. performed the instrumental analysis. A.-T.R. and N.E.M. analysed the data. All authors contributed to writing and editing the manuscript. All authors have read and approved the final version of the manuscript.

Funding: The South African National Research Foundation (NRF) is thanked for grant support to NEM (grant number 107067) and IAD (grant number 95818).

Institutional Review Board Statement: Not applicable.

Informed Consent Statement: Not applicable.

Data Availability Statement: The study design information, MS data, data processing and analyses are reported on and incorporated into the main text as well as reference [5].

Acknowledgments: The University of Johannesburg and the National Research Foundation are thanked for bursary support to ATR.

Conflicts of Interest: The authors declare no conflict of interest.

References

1. Seca, A.M.L.; Pinto, D.C.G.A. Biological potential and medical use of secondary metabolites. *Medicines* **2019**, *6*, 66. [CrossRef]
2. Tiwari, R.; Rana, C.S. Plant secondary metabolites: A review. *Int. J. Eng. Res. Gen. Sci.* **2015**, *3*, 661–670. [CrossRef]
3. Bartolome, A.P.; Villaseñor, I.M.; Yang, W.C. *Bidens pilosa* L. (Asteraceae): Botanical properties, traditional uses, phytochemistry, and pharmacology. *Evid. Based Complement. Altern. Med.* **2013**, *2013*, 340215. [CrossRef] [PubMed]
4. Khoza, B.S.; Gbashi, S.; Steenkamp, P.A.; Njobeh, P.B.; Madala, N.E. Identification of hydroxylcinnamoyl tartaric acid esters in *Bidens pilosa* by UPLC-tandem mass spectrometry. *S. Afr. J. Bot.* **2016**, *103*, 95–100. [CrossRef]
5. Ramabulana, A.-T.; Steenkamp, P.; Madala, N.; Dubery, I.A. Profiling of chlorogenic acids from *Bidens pilosa* and differentiation of closely related positional isomers with the aid of UHPLC-QTOF-MS / MS-based in-source collision-induced dissociation. *Metabolites* **2020**, *10*, 178. [CrossRef]
6. Ong, K.W.; Hsu, A.; Tan, B.K.H. Anti-diabetic and anti-lipidemic effects of chlorogenic acid are mediated by AMPK activation. *Biochem. Pharmacol.* **2013**, *85*, 1341–1351. [CrossRef]
7. Makola, M.M.; Dubery, I.A.; Koorsen, G.; Steenkamp, P.A.; Kabanda, M.M.; Du Preez, L.L.; Madala, N.E. The effect of geometrical isomerism of 3,5-dicaffeoylquinic acid on its binding affinity to HIV-integrase enzyme: A molecular docking study. *Evid. Based Complement. Altern. Med.* **2016**, *2016*, 1–10. [CrossRef]
8. Lodise, O.; Patil, K.; Karshenboym, I.; Prombo, S.; Chukwueke, C.; Pai, S.B. Inhibition of prostate cancer cells by 4,5-dicaffeoylquinic acid through cell cycle arrest. *Prostate Cancer* **2019**, *2019*, 4520645. [CrossRef]
9. Ramirez-Estrada, K.; Vidal-Limon, H.; Hidalgo, D.; Moyano, E.; Golenioswki, M.; Cusidó, R.M.; Palazon, J. Elicitation, an effective strategy for the biotechnological production of bioactive high-added value compounds in plant cell factories. *Molecules* **2016**, *21*, 182. [CrossRef]
10. Marchev, A.S.; Yordanova, Z.P.; Georgiev, M.I. Green (cell) factories for advanced production of plant secondary metabolites. *Crit. Rev. Biotechnol.* **2020**, *40*, 443–458. [CrossRef] [PubMed]
11. Phillips, G.C.; Garda, M. Plant tissue culture media and practices: An overview. *Plant* **2019**, *55*, 242–257. [CrossRef]
12. Gaspar, T.; Keveks, C.; Penel, C.; Greppin, H.; Reid, D.M.; Thorpe, T.A. Plant hormones and plant growth regulators in plant tissue culture. *In Vitro Cell. Dev. Biol. Plant* **1996**, *32*, 272–289. [CrossRef]
13. Moubayidin, L.; Mambro, R.D.; Sabatini, S. Cytokinin-auxin crosstalk. *Trends Plant Sci.* **2009**, *14*, 557–562. [CrossRef] [PubMed]
14. Su, Y.H.; Liu, Y.B.; Zhang, X.S. Auxin-cytokinin interaction regulates meristem development. *Mol. Plant* **2011**, *4*, 616–625. [CrossRef]
15. Lee, Z.H.; Hirakawa, T.; Yamaguchi, N.; Ito, T. The roles of plant hormones and their interactions with regulatory genes in determining meristem activity. *Int. J. Mol. Sci.* **2019**, *20*, 4065. [CrossRef]
16. Coenen, C.; Lomax, T.L. Auxin-cytokinin interactions in higher plants: Old problems and new tools. *Trends Plant Sci.* **1997**, *2*, 351–355. [CrossRef]

17. Shin, J.; Bae, S.; Seo, P.J. De novo shoot organogenesis during plant regeneration. *J. Exp. Bot.* **2020**, *71*, 63–72. [CrossRef] [PubMed]
18. Pernisová, M.; Klíma, P.; Horák, J.; Válková, M.; Malbeck, J.; Souček, P.; Reichman, P.; Hoyerová, K.; Dubová, J.; Friml, J.; et al. Cytokinins modulate auxin-induced organogenesis in plants via regulation of the auxin efflux. *Proc. Natl. Acad. Sci. USA* **2009**, *106*, 3609–3614. [CrossRef]
19. Schaller, G.E.; Bishopp, A.; Kieber, J.J. The yin-yang of hormones: Cytokinin and auxin interactions in plant development. *Plant Cell* **2015**, *27*, 44–63. [CrossRef]
20. Hamany Djande, C.Y.; Steenkamp, P.A.; Steenkamp, A.; Piater, L.A.; Madala, N.E. Habituated *Moringa oleifera* callus retains metabolic responsiveness to external plant growth regulators. *Plant Cell. Tissue Organ Cult.* **2019**. [CrossRef]
21. Malik, S.; Zia, M.; Rehman, R.; Chaudhary, F. In vitro plant regeneration from direct and indirect organogenesis of *Momordica charantia*. *Pakistan J. Biol. Sci.* **2007**, *10*, 4118–4122. [CrossRef] [PubMed]
22. Yu, J.; Liu, W.; Liu, J.; Qin, P.; Xu, L. Auxin control of root organogenesis from callus in tissue culture. *Front. Plant Sci.* **2017**, *8*, 1–4. [CrossRef]
23. Cardoso, J.C.; de Oliveira, M.E.B.S.; de Cardoso, F.I. Advances and challenges on the in vitro production of secondary metabolites from medicinal plants. *Hortic. Bras.* **2019**, *37*, 124–132. [CrossRef]
24. Alagarsamy, K.; Shamala, L.F.; Wei, S. Influence of media supplements on inhibition of oxidative browning and bacterial endophytes of *Camellia sinensis* var. *sinensis*. *3 Biotech* **2018**, *8*, 1–7. [CrossRef]
25. Murata, M.; Nishimura, M.; Murai, N.; Haruta, M.; Itoh, Y. A Transgenic apple callus showing reduced polyphenol oxidase activity and lower browning potential. *Biosci. Biotechnol. Biochem.* **2014**, *65*, 383–388. [CrossRef]
26. Wu, J.; Lin, L. Ultrasound-induced stress responses of *Panax ginseng* Cells: Enzymatic browning and phenolics production. *Biotechnol. Prog.* **2002**, *18*, 862–866. [CrossRef] [PubMed]
27. Chaudhary, G.; Dantu, P.K. Evaluation of callus browning and develop a strategically callus culturing of *Boerhaavia diffusa* L. *J. Plant Dev.* **2015**, *22*, 47–58.
28. Pischke, M.S.; Huttlin, E.L.; Hegeman, A.D.; Sussman, M.R. A Transcriptome-based characterization of habituation in plant tissue culture. *Plant Physiol.* **2006**, *140*, 1255–1278. [CrossRef] [PubMed]
29. Gaspar, T.; Kevers, C.; Bisbis, B.; Franck, T.; Crevecoeur, M.; Greppin, H.; Dommes, J. Special symposium: In vitro plant recalcitrance loss of plant organogenic totipotency in the course of in vitro neoplastic progression. *In Vitro Cell Dev. Biol. Plant* **2000**, *36*, 171–181. [CrossRef]
30. Zhang, T.Q.; Lian, H.; Tang, H.; Dolezal, K.; Zhou, C.M.; Yu, S.; Chen, J.H.; Chen, Q.; Liu, H.; Ljung, K.; et al. An intrinsicmicroRNA timer regulates progressive decline in shoot regenerative capacity in plants. *Plant Cell* **2015**, *27*, 349–360. [CrossRef] [PubMed]
31. Ikeuchi, M.; Ogawa, Y.; Iwase, A.; Sugimoto, K. Plant regeneration: Cellular origins and molecular mechanisms. *Development* **2016**, *143*, 1442–1451. [CrossRef] [PubMed]
32. Sharma, A.; Shahzad, B.; Rehman, A.; Bhardwaj, R.; Landi, M.; Zheng, B. Response of phenylpropanoid pathway and the role of polyphenols in plants under abiotic stress. *Molecules* **2019**, *24*, 2452. [CrossRef]
33. Stojakowska, A.; Malarz, J.; Kiss, A.K. Hydroxycinnamates from elecampane (*Inula helenium* L.) callus culture. *Acta Physiol. Plant.* **2016**, *38*, 1–5. [CrossRef]
34. Ramabulana, A.T.; Steenkamp, P.A.; Madala, N.E.; Dubery, I.A. Profiling of altered metabolomic states in *Bidens pilosa* leaves in response to treatment by methyl jasmonate and methyl salicylate. *Plants* **2020**, *9*, 1275. [CrossRef] [PubMed]
35. Luczkiewicz, M.; Kokotkiewicz, A.; Glod, D. Plant growth regulators affect biosynthesis and accumulation profile of isoflavone phytoestrogens in high-productive in vitro cultures of *Genista tinctoria*. *Plant Cell. Tissue Organ Cult.* **2014**, *118*, 419–429. [CrossRef]
36. Lee, Y.; Lee, D.E.; Lee, H.S.; Kim, S.K.; Lee, W.S.; Kim, S.H.; Kim, M.W. Influence of auxins, cytokinins, and nitrogen on production of rutin from callus and adventitious roots of the white mulberry tree (*Morus alba* L.). *Plant Cell. Tissue Organ Cult.* **2011**, *105*, 9–19. [CrossRef]
37. Jamwal, K.; Bhattacharya, S.; Puri, S. Plant growth regulator mediated consequences of secondary metabolites in medicinal plants journal of applied research on medicinal and aromatic plants plant growth regulator mediated consequences of secondary metabolites in medicinal plants. *J. Appl. Res. Med. Aromat. Plants* **2018**, *9*, 26–38. [CrossRef]
38. Wiklund, S.; Johansson, E.; Sjo, L.; Shockcor, J.P.; Gottfries, J.; Moritz, T.; Trygg, J. Visualization of GC/TOF-MS-based metabolomics data for identification of biochemically interesting compounds using OPLS class models. *Anal. Chem.* **2008**, *80*, 115–122. [CrossRef]
39. Bartel, J.; Krumsiek, J.; Theis, F.J. Statistical methods for the analysis of high-throughput metabolomics data. *Comput. Struct. Biotechnol. J.* **2013**, *4*, e201301009. [CrossRef]
40. Tugizimana, F.; Steenkamp, P.A.; Piater, L.A.; Dubery, I.A. A conversation on data mining strategies in LC-MS untargeted metabolomics: Pre-processing and pre-treatment steps. *Metabolites* **2016**, *6*, 40. [CrossRef]
41. Godzien, J.; Ciborowski, M.; Angulo, S.; Barbas, C. From numbers to a biological sense: How the strategy chosen for metabolomics data treatment may affect final results. A Practical example based on urine fingerprints obtained by LC-MS. *Electrophoresis* **2013**, *34*, 2812–2826. [CrossRef]
42. Granato, D.; Santos, J.S.; Escher, G.B.; Ferreira, B.L.; Maggio, R.M. Use of principal component analysis (PCA) and hierarchical cluster analysis (HCA) for multivariate association between bioactive compounds and functional properties in foods: A critical perspective. *Trends Food Sci. Technol.* **2018**, *72*, 83–90. [CrossRef]

43. Rodriguez, M.Z.; Comin, C.H.; Casanova, D.; Bruno, O.M.; Amancio, D.R.; Costa, L.D.F.; Rodrigues, F.A. Clustering algorithms: A comparative approach. *PLoS ONE* **2019**, *14*, 1–34. [CrossRef]
44. Chong, J.; Soufan, O.; Li, C.; Caraus, I.; Li, S.; Bourque, G.; Wishart, D.S.; Xia, J. MetaboAnalyst 4.0: Towards more transparent and integrative metabolomics analysis. *Nucleic Acids Res.* **2018**, *46*, 486–494. [CrossRef]
45. Park, C.H.; Yeo, H.J.; Park, Y.J.; Morgan, A.M.A.; Arasu, M.V.; Al-Dhabi, N.A.; Park, S.U.; Cravotto, G. Influence of indole-3-acetic acid and gibberellic acid on phenylpropanoid accumulation in common buckwheat (*Fagopyrum esculentum* Moench) sprouts. *Molecules* **2017**, *22*, 374. [CrossRef]
46. Teixeira, J.; Gaspar, A.; Garrido, E.M.; Garrido, J.; Borges, F. Hydroxycinnamic acid antioxidants: An electrochemical overview. *Biomed. Res. Int.* **2013**, *2013*, 251754. [CrossRef] [PubMed]
47. Hassan, A.M.; Mohamed, H.E.; Mostafa, E.M. Nutrient starvation enhances the phenolic compounds and antioxidant activity in *Azolla caroliniana* plant. *Egypt. J. Bot.* **2020**, *60*, 239–247. [CrossRef]
48. Murashige, T.; Skoog, F. A Revised medium for rapid growth and bio assays with tobacco tissue cultures. *Physiol. Plant.* **1962**, *15*, 474–497. [CrossRef]
49. Van der Berg, R.A.; Hoofsloot, H.C.; Westerhuis, J.A.; Smilde, A.K.; van der Werf, M. Centering, scaling, and transformations: Improving the biological information content of metabolomics data. *Cancer Res.* **2006**, *7*, 1–15. [CrossRef]
50. Cheng, J.; Lan, W.; Zheng, G.; Gao, X. Metabolomics: A high-throughput platform for metabolite profile exploration. *Methods Mol. Biol.* **2018**, *1754*, 265–292. [CrossRef]
51. Sumner, L.W.; Samuel, T.; Noble, R.; Gmbh, S.D.; Barrett, D.; Beale, M.H.; Hardy, N. Proposed minimum reporting standards for chemical analysis Chemical Analysis Working Group (CAWG) Metabolomics Standards Initiative (MSI). *Metabolomics* **2007**, *3*, 211–221. [CrossRef] [PubMed]
52. Dictionary of Natural Products. Available online: http://dnp.chemnetbase.com/faces/chemical/ChemicalSearch.xhtml (accessed on 20 July 2020).
53. Kyoto Encyclopedia of Genes and Genomes. Available online: https://www.genome.jp/kegg/ (accessed on 20 July 2020).
54. Chemspider Chemspider. Available online: http://www.chemspider.com (accessed on 20 July 2020).
55. MetaboAnalyst MetaboAnalyst. Available online: https://www.metaboanalyst.ca/ (accessed on 15 August 2020).
56. Xia, J.; Psychogios, N.; Young, N.; Wishart, D.S. MetaboAnalyst: A web server for metabolomic data analysis and interpretation. *Nucleic Acids Res.* **2009**, *37*, 652–660. [CrossRef] [PubMed]

plants

MDPI

Article

Gentianella lutescens subsp. *carpatica* J. Holub.: Shoot Propagation In Vitro and Effect of Sucrose and Elicitors on Xanthones Production

Dijana Krstić-Milošević [1,*], Nevena Banjac [1], Teodora Janković [2], Dragan Vinterhalter [1] and Branka Vinterhalter [1]

1 Department of Plant Physiology, Institute for Biological Research "Siniša Stanković"—National Institute of the Republic of Serbia, University of Belgrade, Bulevar despota Stefana 142, 11000 Belgrade, Serbia; mitic.nevena@ibiss.bg.ac.rs (N.B.); dvinterhalter@yahoo.com (D.V.); horvat@ibiss.bg.ac.rs (B.V.)

2 Insitute for Medicinal Plants Research "Dr Josif Pančić", Tadeuša Košćuška 1, 11000 Belgrade, Serbia; tjankovic@mocbilja.rs

* Correspondence: dijana@ibiss.bg.ac.rs

Abstract: In vitro shoot culture of the endangered medicinal plant *Gentianella lutescens* was established from epicotyl explants cultured on MS basal medium with 0.2 mg L^{-1} 6-benzylaminopurine (BA) and evaluated for xanthones content for the first time. Five shoot lines were obtained and no significant variations in multiplication rate, shoot elongation, and xanthones profile were found among them. The highest rooting rate (33.3%) was achieved by shoots treated for 2 days with 5 mg L^{-1} indole-3-butyric acid (IBA) followed by cultivation in liquid PGR-free ½ MS medium for 60 days. HPLC analysis revealed the lower content of xanthones—mangiferin, bellidifolin, demethylbellidifolin, demethylbellidifolin-8-*O*-glucoside and bellidifolin-8-*O*-glucoside—in in vitro cultured shoots compared to wild growing plants. The increasing concentration of sucrose, sorbitol and abiotic elicitors salicylic acid (SA), jasmonic acid (JA) and methyl jasmonate (MeJA) altered shoot growth and xanthone production. Sucrose and sorbitol applied at the highest concentration of 233.6 mM increased dry matter percentage, while SA at 100 μM promoted shoot growth 2-fold. The increased sucrose concentration enhanced accumulation of xanthones in shoot cultures 2–3-fold compared to the control shoots. Elicitors at 100–300 μM increased the accumulation of mangiferin, demethylbellidifolin-8-*O*-glucoside, and bellidifolin-8-*O*-glucoside almost equally, while MeJA at the highest concentration of 500 μM enhanced amount of aglycones demethylbellidifolin and bellidifolin 7-fold compared to the control. The obtained results facilitate conservation of *G. lutescens* and pave the way for further research on large-scale shoot propagation and production of pharmacologically active xanthones.

Keywords: shoot culture; secondary metabolites; HPLC; bellidifolin; osmotic stress

Citation: Krstić-Milošević, D.; Banjac, N.; Janković, T.; Vinterhalter, D.; Vinterhalter, B. *Gentianella lutescens* subsp. *carpatica* J. Holub.: Shoot Propagation In Vitro and Effect of Sucrose and Elicitors on Xanthones Production. *Plants* **2021**, *10*, 1651. https://doi.org/10.3390/plants10081651

Academic Editor: Iyyakkannu Sivanesan

Received: 28 June 2021
Accepted: 6 August 2021
Published: 11 August 2021

1. Introduction

The genus *Gentianella* Moench (Gentianeceae) encompasses about 250 species growing mainly in temperate or mountain habitats in Europe, South America, New Zealand, and Australia [1]. In Europe, the genus *Gentianella* consists of 22 species with numerous subspecies and taxa that are distributed in the Alps, Carpathians, in the Tatra mountains, as well as in the mountains of the Balkan Peninsula [2]. In Serbia, the genus *Gentianella* is represented by six species: *G. austriaca*, *G. bulgarica*, *G. axilaris*, *G. ciliata*, *G. praecox*, and *G. crispata* [3].

Apart their ornamental value, *Gentianella* species have been well-known as traditional medicines since ancient times. In South America they have been used as a traditional remedy for the treatment of digestive and liver problems [4,5]. *Gentianella amarella* is known in the traditional medicine of Mongolia to cure headache, hepatitis, fever, and gallbladder disorders [6]. Like other species of the Gentianaceae family, *Gentianella* plants are characterized by the universal occurrence of three main groups of secondary metabolites, iridoids, flavonoids, and xanthones [7].

Naturally occurring xanthones have been attracting attention for a long time due to their specific bioactivities and occupy an important position in the pharmacology and chemistry of natural products. Xanthone compounds typical of *Gentianella* species belong to the bellidifolin type of xanthones, which mostly occur in the form of *O*-glycosides. They are responsible for wide range of therapeutic properties attributed to *Gentianella* plants. The xanthones bellidifolin and demethylbellidifolin, the principal constituents of many gentianellas, have been reported to show cardioprotective effects, antioxidant, antimicrobial, and antidiabetic activity, as well as displaying significant potential to inhibit acetylcholinesterase and monoamino oxidase activity [8]. Bellidifolin exhibited a variety of pharmacological activities, including prominent hypoglycemic [9] and neuroprotective activities [10]. These findings support the potential use of xanthone compounds as new drugs in treating aging-related neurodegenerative disorders [11] and also as useful candidates for therapy of type 2 diabetes [9].

Although finding *Gentianella* species as a reach source of important phytochemicals has heightened research interest in these species, many of them have not yet been properly investigated due to the low availability of plant material. A large number of *Gentianella* species are rare and endemic or grow in inaccessible localities, while the others became endangered from excessive harvesting or adverse environmental conditions. Such an unfavorable situation is the case with the endemic species *G. lutescens* subsp. *carpatica*, which we discovered in Serbia for the first time during field research on Povlen mountain (Figure 1). A survey of literature indicated that this species has not been phytochemically investigated so far and very scarce data can be found about it.

Figure 1. *Gentianella lutescens* subsp. *carpatica* in a natural habitat on Povlen mountain (locality Razbojište), Serbia.

It is biennial 3–40-cm-tall plant, and is simple or branched in the upper part, forming a racemose inflorescence panicle-like, umbrella-shape with reddish to violet flowers. *Gentianella lutescens* subsp. *carpatica* was located in Eastern and Central Europe and northern part of the Balkan Peninsula, mainly in the mountains, with ecotypic variants in Austria, Bulgaria, Czech, Germany, Yugoslavia, and Poland [12]. This species was reported as relatively common in the Czech Republic before 1950; however, nowadays it is considered a critically threatened plant surviving on only a few sites [13]. The fact that the extant populations are small, often less than ten individuals, signify that *G. lutescens* is a critically endangered plant species that is included in both the European and world red list, according to International Union for Conservation of Nature's (IUCN) Red List [14].

Advances in plant biotechnology and tissue culture have proven to be a valuable tool for large-scale propagation, storage, reintroduction, as well as the production of secondary metabolites of endangered medicinal plant species [15–17].

Many papers have been published over the last two decades related to in vitro culture studies of more eminent members of Gentianaceae, such as *Gentiana* spp. and *Swertia* spp. However, these studies were performed in only a few species of the genus *Gentianella*, and included *G. austriaca* [18], *G. bulgarica* [19,20], *Gentianella albifpra* [21], and *G. bicolor* [22]. Considering the rarity and medicinal importance of *G. lutescens*, tissue culture would be a suitable tool, not only for conservation purposes, but also for overcoming the deficit of plant material in order to perform phytochemical investigations and in the development of technology for the extraction of important metabolites [23]. Although a lower content of secondary metabolites of interest in the tissue of cultured plants compared to natural plants may limit the applicability of tissue culture [17], it also offers the possibility to stimulate their production using various strategies.

Elicitation with chemical compounds is an effective strategy for improving the production of secondary metabolites in plant cells and tissue cultures. Elicitor molecules trigger signaling events leading to the activation of defense-related genes involved in the biosynthesis of secondary metabolites. Increased transcription of genes encoding biosynthetic enzymes of central phenylpropanoid pathway (phenyilalanine-ammonia lyase (PAL) and 4-coumarate-CoA-ligase (4 CL)) and specific enzyme benzophenone synthase (BPS) of the xanthone synthesis branch route has been demonstrated upon elicitation of *Hypericum perforatum* cell cultures [24]. Salicylic acid (SA), jasmonic acid (JA), and methyl jasmonate (MeJA) are plant defense signaling molecules, and, when exogenously added, they induce systemic acquired resistance against stress, thus acting as elicitors. Several studies reported higher production of xanthones and flavonoids in cell suspensions of *H. perforatum* treated with SA, JA, or MeJA [25–27]. Exposure to high levels of sucrose or sorbitol have also induced an increased accumulation of some secondary metabolites caused by osmotic stress [28,29].

In view of medicinal and conservation importance of *G. lutescens*, the aim of the current work was to evaluate in vitro grown shoot culture of *G. lutescens* as an alternative, sustainable, and stable source of xanthones. In order to increase biomass and xanthones content in *G. lutescens* shoots, the effect of sucrose, sorbitol, and elicitors JA, MeJA, and SA, on shoot growth and xanthones production was investigated.

2. Results and Discussion

2.1. HPLC-DAD Analysis of Secondary Metabolites of Wild Grown G. lutescens Plants

The chemical profile of methanol extract of aerial parts of wild-growing *G. lutescens* plants analyzed using the HPLC-DAD technique is presented in Figure 2. Similarly, as in other *Gentianella* species, three groups of secondary metabolites, secoiridoids, flavone-C-glucosides, and xanthones, were detected. Chromatographic analysis identified the presence of swertiamarin and gentiopicrin (peaks 1 and 2, respectively) as the most common secoiridoid compounds, which appeared to be present in all species of Gentianaceae [30].

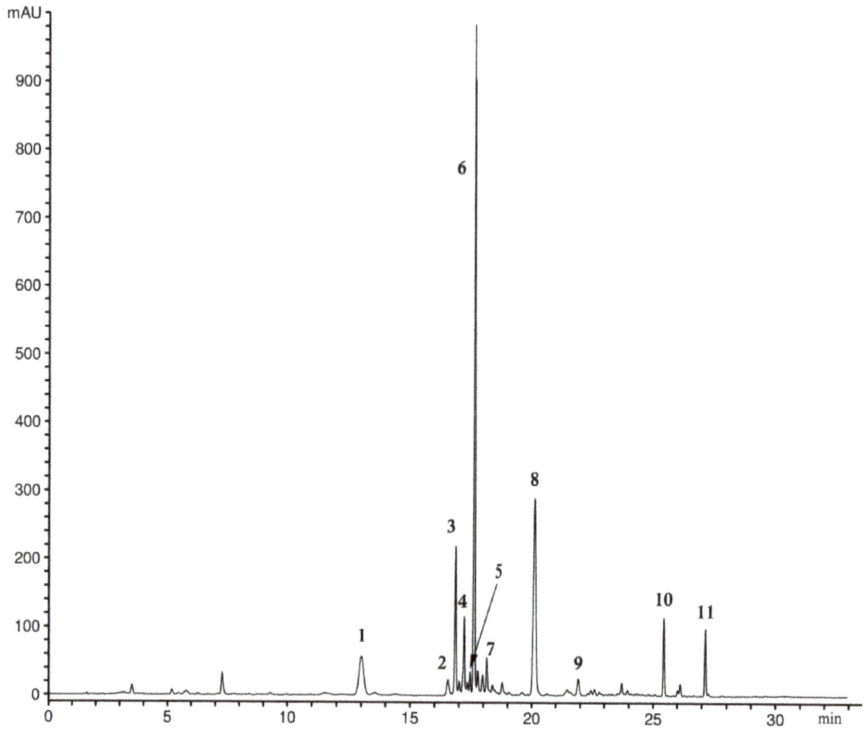

Figure 2. HPLC profile (λ = 260 nm) of methanol extract of *G. lutescens*. Peaks: 1—swertiamarin, 2—gentiopicrin, 3—mangiferin, 4—campestroside, 5—isoorientin, 6—demethylbellidifolin-8-*O*-glucoside, 7—swertisin, 8—bellidifolin-8-*O*-glucoside, 9—veratriloside, 10—demethylbellidifolin, 11—bellidifolin.

Peaks 5 and 7 were identified as isoorientin and swertisin, the most represented *C*-glucoflavones in the *Gentianella* species. The seven remaining peaks presented in the chromatogram were detected as xanthone compounds. The precise identification of each xanthone was confirmed using the HPLC co-injection method using reference xanthone standards, previously isolated in our laboratory [31]. The two dominant peaks (Figure 2) belong to demethylbellidifolin-8-*O*-glucoside (6) and bellidifolin-8-*O*-glucoside (8), xanthones with 1,3,5,8-oxidation pattern characteristic for *Gentianella* species. HPLC also revealed the presence of a tetrahydroxanthone glucoside named campestroside (4), a partially saturated analog of demethylbellidifolin-8-*O*-glucoside. A xanthone with such a structure need special attention since it is rare in nature and its occurrence is of particular chemotaxonomic and biogenetic significance [31]. Compound 9 was identified as xanthone-*O*-glucoside veratriloside, and this compound was reported to be the first 1,3,4,7-oxygenated xanthone isolated from the genus *Gentianella* [20,31]. Peak 3 belongs to *C*-glucoxanthone mangiferin, one of the well-known naturally occurring xanthones, widespread among angiosperms. The occurrence of mangiferin together with flavone-*C*-glucosides isoorientin and swertisin is most common and is typical for *Gentianella* species. The peaks detected at the end of the chromatogram were identified as xanthone aglycons demethylbellidifolin (10) and bellidifolin (11). Figure 3 shows the chemical structures of the secoiridoid and xanthone compounds identified in *G. lutescens*.

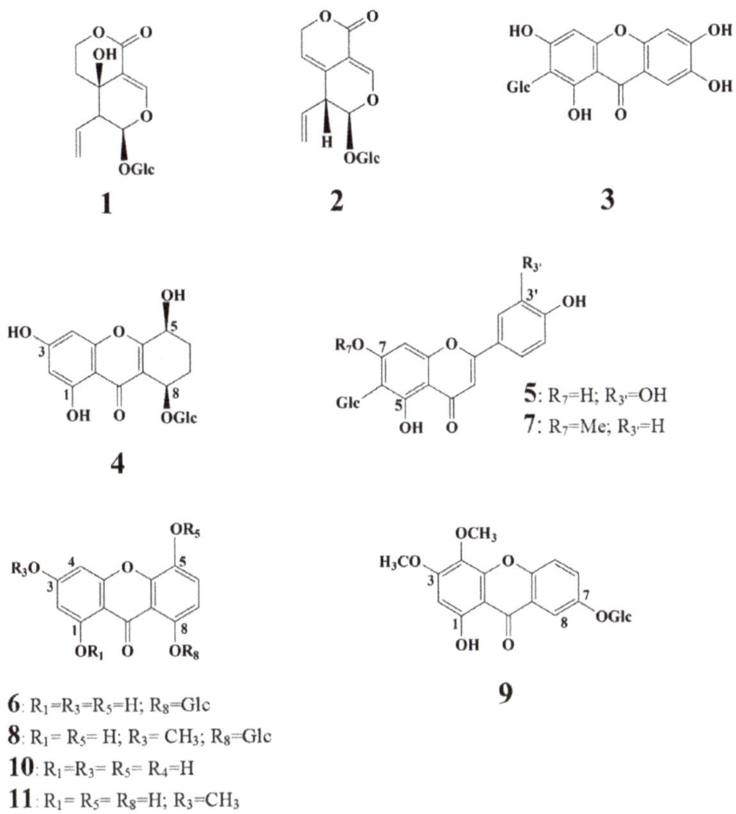

Figure 3. Chemical structures of secoiridoid and xanthone compounds identified in *G. lutescens*. **1**—swertiamarin, **2**—gentiopicrin, **3**—mangiferin, **4**—campestroside, **5**—isoorientin, **6**—demethylbellidifolin-8-*O*-glucoside, **7**—swertisin, **8**—bellidifolin-8-*O*-glucoside, **9**—veratriloside, **10**—demethylbellidifolin, **11**—bellidifolin.

Considering that xanthones are becoming increasingly important compounds that possess a broad spectrum of biological and pharmacological activities, in this study we analyzed secondary metabolites from *G. lutescens* cultured in vitro. Specifically, we focused on the chemical analysis of five dominant xanthones: mangiferin, demethylbellidifolin-8-*O*-glucoside (DMB-8-*O*-glc), bellidifolin-8-*O*-glucoside (bell-8-*O*-glc), and the aglycons demethylbellidifolin (DMB) and bellidifolin.

2.2. *In Vitro Shoot Propagation of* G. lutescens

Immature seeds of *G. lutescens* (Figure 4A) were germinated for 10 days at a rate of 5%, and non-contaminated seeds were found. The maximum response of the epicotyl explants to produce new shoots was observed after 35 days of culture on shoot induction medium (Figure 4B). The genotype of the individual seedlings did not have a significant effect on the shoot proliferation response, as a multiplication index of about 3 was recorded in all five lines. However, the genotype significantly affected the elongation of the main shoot, ranging from 14.49 mm in line 1 to 23.05 mm in line 3 (Table 1).

Figure 4. In vitro propagation of *G. lutescens*. (**A**) Open pod with immature seeds (*bar* = 5 mm), (**B**) shoot multiplication on BM + 0.2 mg L^{-1} BA (*bar* = 10 mm), (**C**) root elongation in PGR-free ½ MS liquid medium after treatment of shoots with 5.0 mg L^{-1} IBA for 2 days (*bar* = 10 mm).

Table 1. Shoot induction and multiplication and the length of the main shoot in *G. lutescens*. Epicotyl explants of five seedling lines were cultivated on BM + 0.2 mg L^{-1} BA for 35 days. Values represent the means ± SE from two experiments with 35–40 samples per line (n = 75–80). Data were analyzed by one-way ANOVA analysis. Within each column means followed by different letters are significantly different according to Fisher's LSD test at $p \leq 0.05$. Multiplication index: main shoot + axillary buds.

Line (BA 0.2 mg L^{-1})	No. of Explants	Multiplication Index ± SE	Length of Main Shoot (mm) ± SE	
line 1	80	3.03 ± 0.23 ab	14.49 ± 0.65 a	
line 2	77	3.31 ± 0.19 ab	17.40 ± 0.92 ab	
line 3	80	3.53 ± 0.23 b	23.05 ± 1.67 c	
line 4	75	2.76 ± 0.25 a	17.19 ± 1.22 ab	
line 5	80	3.41 ± 0.26 ab	19.39 ± 1.07 b	
ANOVA Source of Variation	**Df**	**Mean Square**	***F*-Ratio**	***p*-Value**
Multiplication index	4	7.48476	1.73	0.1428
Length of main shoot	4	804.293	7.67	0.0000

Elongated shoots (≥15 mm) sporadically formed flower buds. The shoots with intense blue–violet flowers developed normally on the cytokinin-containing medium. With the aim to enhance shoot proliferation, the individual shoots of line 5 were transferred onto BM with increasing concentrations of BA (0–2.0 mg L^{-1}). Line 5 was chosen for the shoot proliferation experiment because it showed growth and multiplication stability over time and produced a large number of healthy-looking shoots that were needed to start the experiment. According to ANOVA, shoot multiplication and shoot length were significantly affected by the concentration of BA (Table 2). A higher BA concentration (above 0.5 mg L^{-1}), not only contributed to the enhancement of shoot multiplication rate, but also increased the percent of low-quality vitrified shoots (Table 2). Moreover, the length of the main shoot was significantly decreased with increasing BA concentration. The highest multiplication rate, close to 4, was achieved on basal medium supplemented with 0.2 mg L^{-1} BA along with minimum reduction of the main shoot length compared to cytokinin-free medium. In light of this finding, and considering the reasonable percent of vitrified shoots (7.75%), the above medium formulation was selected as optimal for shoot culture multiplication and maintenance. Unexpectedly, a relatively satisfactory multiplication rate (2.89) was achieved on cytokinin-free medium relative to BA-containing media. This can be explained by prolonged effect of BA from shoot induction medium.

Table 2. Effect of increasing concentrations (0–2.0 mg L^{-1}) of BA on shoot multiplication, the length of the main shoot and vitrification incidence of *G. lutescens* line 5. Values represent the means ± SE from 3–4 experiments with 25–40 samples per treatment (n = 80–129). Data were analyzed by one-way ANOVA analysis. Within each column means followed by different letters are significantly different according to Fisher's LSD test at $p \leq 0.05$. Results were scored after 35 days. Multiplication index—main shoot + axillary buds.

BA (mg L^{-1})	No. of Expl.	Multiplication Index ± SE	Length of Main Shoot (mm) ± SE	Vitrification	
				No.	%
0	85	2.89 ± 0.16 a	15.02 ± 0.66	1	0.8
0.05	100	3.16 ± 0.17 a	15.85 ± 0.50 e	1	1.0
0.1	128	3.20 ± 0.14 a	13.66 ± 0.46 cd	7	5.47
0.2	90	3.92 ± 0.25 b	15.48 ± 0.64 cd	10	7.75
0.5	129	3.22 ± 0.17 a	13.09 ± 0.48 bc	24	18.6
1.0	128	2.88 ± 0.15 a	12.09 ± 0.38 b	20	15.6
2.0	80	2.94 ± 0.20 a	10.46 ± 0.47 a	14	17.5
ANOVA Source of Variation	**Df**	**Mean Square**	***F*-Ratio**	***p*-Value**	
Multiplication index	6	12.2028	3.72	0.0012	
Length of main shoot	6	351.866	12.96	0.0000	

In cultures maintained on basal medium supplemented with 0.2 mg L^{-1} BA more than one year, the number of shoots with spontaneously developed flowering buds gradually decreased over time and became rare in those maintained more than 5 years. Cytokinin was a common requirement for in vitro flowering, which also occurred in *G. austriaca* [18] and *G. bulgarica* shoot cultures [19]. However, while precocious in vitro flowering in these two species threatened shoot multiplication at a higher rate due to decay of the main shoot after flowering, it appeared sporadically in *G. lutescens* and has not been observed to significantly affect multiplication. Nevertheless, in vitro flowering could serve as an important tool for many purposes. Manipulation of different variables in in vitro conditions offers a unique system for studying molecular basis and hormonal regulation of flowering. On the other hand, in vitro flowering could be a valuable tool for the faster release of new varieties [32], enabling also the recombination of genetic material via in vitro fertilization in otherwise non-hybridizable lines. Since the multifunctional usability of in vitro flowering, especially in rare and endangered plant species, such as *G. lutescens*, comprehensive study of this phenomenon could be the main goal of further research studies in *G. lutescens*.

To the best of our knowledge, this is the first report on the establishment of shoot cultures of endemic *G. lutescens*. In general, *Gentianella* species were, so far, barely investigated compared to their closest relatives belonging to the genus *Gentiana*. Huo and Zheng [21] were the first to reported shoot regeneration from calli of *G. albifpra* cultured on medium with 3.0 mg L^{-1} 2,4-D + 1.0 mg L^{-1} KIN. Later on, in vitro propagation was achieved from epicotyls of sterile germinated mature seeds of *G. austriaca* [18] and *G. bulgarica* [19], which are endemic in central part of the Balkan Peninsula. Incorporation of cytokinin promoted shoot explant proliferation in *Gentianella* species and BA was found to be superior for new shoot formation. BA was also effective for inducing shoot proliferation in the previously mentioned *Gentianella* species, *G. austriaca* and *G. bulgarica*, where BA at 0.2–0.5 mg L^{-1} (depending on species) with the addition 0.1 mg L^{-1} NAA was applied [18,19]. More recently, Solorzano et al. [22] reported shoot regeneration from leaf-explant-derived calli of *G. bicolor* on medium containing a combination of KIN and 2,4-D. However, shoot tips, epicotyl, and nodal segments are generally preferred explants for multiplication of most plant species due to the presence of pre-existing meristems. They can be easily developed into shoots that ensure clonal fidelity [33]. Genetic fidelity and true-to-type regenerated plants are very important for both the germplasm maintenance for plant conservation purposes and mass shoot propagation that ensures a continuous supply of uniform genetic plant material for large-scale secondary metabolite production [34,35].

Rooting of *G. lutescens* shoots was performed on a medium with a reduced content of MS mineral salts (by half (½ MS)), supplemented with increasing concentrations of IBA at 0.2–5.0 mg L^{-1} (Table 3). However, if only solid medium was used, auxin-induced root primordia did not elongate and redundant calli were observed, especially at IBA 0.5–2.0 mg L^{-1} (data not shown). Transfer of shoots with induced roots to solid auksin-free medium, with or without active charcoal (1%) or GA_3 (0.1 mg L^{-1}), also did not stimulate elongation of root primordia (data not shown). Elongation of IBA-induced roots was achieved by transferring shoots into liquid PGR-free ½ MS medium. Decades of work with gentians and other plant species suggest that increased concentrations of auxin should reduce the length of induction treatment. Otherwise calli will form instead of roots. Hence, in *G. lutescens*, we used different root induction treatments corresponding to auxin concentration (Table 3). Elongation of the roots in the liquid medium was very slow and required about 60 days. The best rooting was achieved by induction with IBA 5.0 mg L^{-1} for 2 days, followed by root elongation in liquid PGR-free ½ MS medium for 61 days. This protocol provided the highest percentage of rooted shoots (33%), number of shoots with induced root primordia (23.3), number of roots per rooted shoots (5.4), and root elongation (11.2 mm) (Table 3; Figure 4C). These results indicated *G. lutescens* as a highly recalcitrant species in terms of rooting potential that should be further improved to obtain enough healthy and functional plants for successful acclimatization and ex situ and in situ conservation. Other *Gentianella* species also displayed restricted rooting potential. The highest rooting percentage was obtained for *G. austriaca*, where 47.3% of shoots formed roots on solid MS medium with 4.92 μM (1.0 mg L^{-1}) IBA [18]. However, only 1–2% shoots of *G. bulgarica* spontaneously rooted on plant growth regulator free medium [19].

2.3. HPLC-DAD Analysis of Secondary Metabolites of *G. lutescens* Cultured In Vitro

HPLC-DAD analysis of methanol extracts obtained from five shoot lines of *G. lutescens* showed no significant differences between them in either qualitative or in quantitative xanthone composition. It can be noticed that all 5 lines produced a lower content of xanthones compared to the plant material collected in nature (Figure 5). This was not surprising result since the relatively low number of in vitro plants surpassed wild plants in terms of the amounts of secondary metabolites [36].

According to ANOVA, cytokinin BA, necessary for shoot multiplication of *G. lutescens*, did not have a significant effect on the production of xanthones (Supplementary material, Figure S1). However, a different effect of BA was reported in *G. bulgarica* shoots where xanthone production was strongly affected by BA. Namely, it has been shown that the content of xanthones linearly increased with increasing BA concentration (0.2–1.0 mg L^{-1}) in medium [19]. The stimulatory effect of BA on secondary metabolite production has been shown in shoot cultures of *G. austriaca* (2.2–4.4 μM) [18] and *Gentiana asclepiadea* (4.4 μM) [37] as well.

2.4. The Effect of Sucrose and Sorbitol on Shoot Growth and Xanthone Production

According to ANOVA, increasing concentration of sucrose in the growth medium significantly affected shoot growth, dry matter, and flowering (Supplementary Material, Table S1). Thus, additional sucrose, from 58.4 mM to 233.6 mM, gradually decreased shoot fresh weight and growth index (Figure 6A,C). At the same time, the dry weight of the shoots was not significantly affected, while dry matter percentage even significantly increased with rising sucrose concentrations (Figure 6B,D).

Table 3. Effect of pretreatments with IBA on the rooting of shoots of *G. lutescens* line 5. After IBA treatment the shoots were cultivated in liquid PGR-free ½ MS medium with 2% sucrose and fresh medium was added every 7 days. Experiments were repeated 2 times with 15–26 shoots per treatment (n = 30–52). Within each group of experiments (distinct color) means followed by different letters in the column are significantly different according to Fisher's LSD test at $p \leq 0.05$. SE—standard error.

IBA (mg L^{-1})	Auxin Treatment + PGR-Free Liquid Medium = Duration of Experiment (Day)	Explants No.	Shoot with Root Primordia (%)	Rooting (%)	Roots Per Rooted Explant ± SE	Length of the Longest Root (mm) ± SE
0.2	50 + 55 = 105	40	12.5	17.5	2.57 ± 0.84 a	7.29 ± 1.49 a
0.5	50 + 55 = 105	52	13.5	7.7	2.25 ± 0.48 a	8.75 ± 2.56 a
1.0	50 + 55 = 105	50	16.0	14.0	3.43 ± 0.95 a	5.86 ± 2.41 a
2.0	50 + 55 = 105	50	6.0	6.0	2.33 ± 0.33 a	10.51 ± 7.51 a
1.0	14 + 66 = 80	34	8.8	0	-	-
1.0	21 + 59 = 80	37	8.1	3.3	2.0 ± 0	2.0 ± 0
1.0	28 + 52 = 80	45	13.3	0	-	-
2.0	14 + 66 = 80	30	16.7	6.7	4.5 ± 0.5 a	14.0 ± 2.0 a
2.0	21 + 59 = 80	30	16.7	13.3	5.25 ± 1.1 a	13.75 ± 4.27 a
2.0	28 + 52 = 80	33	6.1	0	-	-
5.0	1 + 61 = 62	30	13.3	30.0	2.89 ± 0.5 a	12.4 ± 2.02 a
5.0	2 + 61 = 63	30	23.3	33.3	5.4 ± 1.1 b	11.2 ± 2.6 a
5.0	4 + 61 = 65	30	16.7	33.3	3.5 ± 0.6 ab	7.7 ± 1.23 a
5.0	6 + 59 = 65	30	10.0	16.7	2.4 ± 1.2 a	7.6 ± 1.21 a

In plant tissue culture, sucrose serves as a carbon and energy source necessary for cell division and differentiation [38] and for the regulation of osmotic potential [39]. In the present study, the highest shoot fresh weight (1600 mg) and the growth index (3.2) were recorded in the control *G. lutescens* shoots grown at the lowest (58.4 mM) sucrose. According to Grattapaglia and Machado [40], sugar concentrations lower or higher than 58.42 mM can cause chlorosis or explant deterioration, respectively, in in vitro cultures. The increased concentration of sucrose induces osmotic stress due to enhanced osmotic potential in the growth medium, which inhibits shoots water and nutrients uptake from the medium [41]. This in turn significantly reduces shoots fresh weight and the growth index as compared to the dry weight. Osmotic potential may interfere with nutrient abortion by cells, which is essential to growth and cell division in the aerial parts [42]. This can partly explain the observations of reduced *G. lutescens* shoot fresh weights with sucrose concentrations increasing above 58.4 mM (Figure 6D). Increment of flowering was found (Figure 7) from 0 to 1.4 flowers per Erlenmeyer flask with increasing sucrose concentration from 58.4 in control to 175.2 mM, respectively, and then gradually decreased at 175.2 mM and 233.6 mM sucrose (1.2 and 0.4 flowers per Erlenmeyer flask, respectively) (data not shown) and could be also stress related phenomenon [43].

Additionally, the higher sucrose concentrations at 175.2 and 233.6 mM caused a color change of *G. lutescens* shoots from green to yellow–brown along with stunted growth (Figure 7). High sugar concentrations (45 and 60 g L^{-1}) were found to inhibit the growth of aerial plant parts [44], with a reduction of photosynthetic pigments content [45] compared to plants grown on medium without sucrose, as was verified on *B. zebrina* shoots [46]. The reduction in chlorophyll content in in vitro plants may reduce photosynthetic ability by decreasing light absorption [47].

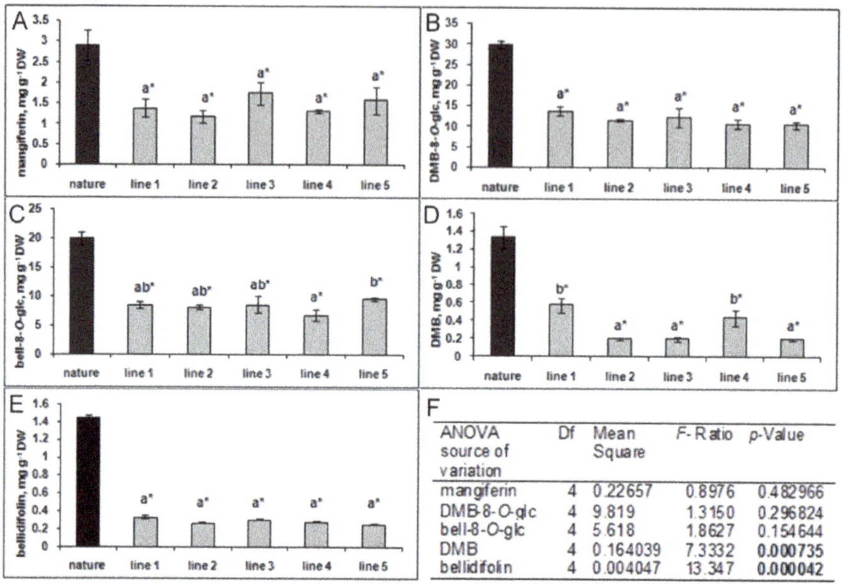

ANOVA source of variation	Df	Mean Square	F-Ratio	p-Value
mangiferin	4	0.22657	0.8976	0.482966
DMB-8-O-glc	4	9.819	1.3150	0.296824
bell-8-O-glc	4	5.618	1.8627	0.154644
DMB	4	0.164039	7.3332	**0.000735**
bellidifolin	4	0.004047	13.347	**0.000042**

Figure 5. The content of xanthones mangiferin (**A**), demethylbellidifolin-8-*O*-glucoside (DMB-8-*O*-glc) (**B**), bellidifolin-8-*O*-glucoside (bell-8-*O*-glc) (**C**), demethylbellidifolin (DMB) (**D**) bellidifolin (**E**) in five shoot lines of *G. lutescens* cultured on BM + 0.2 BA mg L^{-1} for 35 days. Wild *G. lutescens* herb grown in nature was used as a control. Values are the means ± SE of four to six biological replicates (n = 4–6). Data of xanthones content in shoot lines were analyzed by one-way ANOVA (**F**). The values followed by different letters were significantly different according to Fisher's LSD test at $p \leq 0.05$; asterisk (*) indicate a significant difference of values between in vitro shoot lines and wild plants samples according to Student's t-test at $p \leq 0.05$.

On the other hand, although the increased sucrose concentration in the medium above 58.4 mM was not optimal for *G. lutescens* shoot fresh weight and growth index, continuing increasing of dry matter percentage *G. lutescens* shoots over all higher ranges of sucrose tested was indicative (Figure 6). A positive correlation between sucrose in the medium and dry matter content in in vitro plants has been previously reported. Increased availability of sugars in heterotrophic systems has been shown to increase cellulose synthesis, which was correlated to increase in dry weight [48]. In addition, sucrose cleavage in the medium results in glucose and fructose production. It may accelerate cell division and consequently increase the explant weight and volume [49].

Numerous studies indicated sorbitol as the best carbon source for plant multiplication in many species [50,51], as well as to distinguish nutritive effect of enhanced sucrose in the growth medium from its osmotic effect. Here, we analyzed the relative impact of increased sorbitol concentrations on *G. lutescens* shoot cultures. According to ANOVA, increasing concentration of sorbitol significantly altered shoot fresh weight, growth index, dry matter percentage, but not dry weight (Supplementary Material, Table S1).

Results of the present study revealed analogical effect of increasing concentrations of sorbitol (58.4–233.6 mM) with that of sucrose on *G. lutescens* shoot fresh weight, dry weight, and growth index (Figure 6). Moreover, the effect of enhanced sorbitol on increased dry matter percentage (Figure 6H) at an almost identical range compared to that of sucrose (Figure 6D) was observed.

Two-way ANOVA indicated that the type of carbohydrate added did not affect the growth of the shoots, but in vitro flowering and dry matter percentage were significantly affected. However, the concentrations of carbohydrates significantly affected all traits in the same trend (Supplementary material, Table S2).

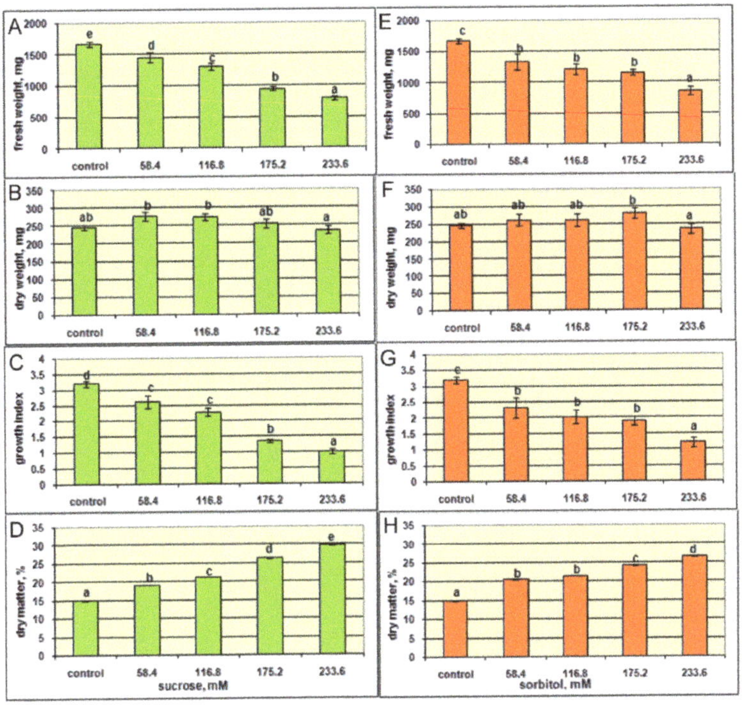

Figure 6. The effect of increasing concentrations (58.4, 161.8, 175.2, and 233.6 mM) of sucrose (**A–D**) and sorbitol (**E–H**) on the growth of shoot cultures of *G. lutescens* (line 5) after 35 days of culture. The increasing concentrations of sucrose or sorbitol were added into control medium (BM + 0.2 mg L^{-1} BA containing 58.4 mM sucrose). Values are the means ± SE of eight to ten biological replicates (n = 8–10). Data were analyzed by one-way ANOVA. Values followed by different letters are significantly different according to Fisher's LSD test at $p \leq 0.05$.

Figure 7. The effect of sucrose at increasing concentrations (58.4, 161.8, 175.2, and 233.6 mM) on the growth and flowering of shoot cultures of *G. lutescens* (line 5) after 35 days of culture (*bar* = 20 mm). The increasing concentrations of sucrose were added into control medium (BM + 0.2 mg L^{-1} BA containing 58.4 mM sucrose). Arrows indicate floral bud development. Right—shoot with normally developed flower regenerated on the medium with 175.2 mM sucrose (*bar* = 5 mm).

These findings strongly confirmed osmotic effect of sucrose on *G. lutescens* and appointed 58.4 mM sucrose as optimal for shoots growth in vitro.

Regarding the effect of sucrose and sorbitol on the xanthone contents in *G. lutescens* shoots, the obtained results showed that increased sucrose concentration stimulated the production of xanthones, whereas the effect of sorbitol was weak or even absent (Figure 8, Supplementary Material, Table S3). Namely, the level of either xanthone analyzed significantly increased with increasing sucrose in the medium and the highest content of xanthones was recorded at the higher sucrose concentration applied. Thus, the content of mangiferin at 233.6 mM of sucrose was more than 3-fold higher compared to shoots cultured on the control medium containing 58.4 mM of sucrose. The highest sucrose concentration enhanced the production of xanthone glucosides DMB-8-*O*-glc and bell-8-*O*-glc for 2.8- and 1.8-times, respectively. The amount of aglycon bellidifolin increased 2.8-times, while the accumulation of DMB increased more than 4 times compared to the control.

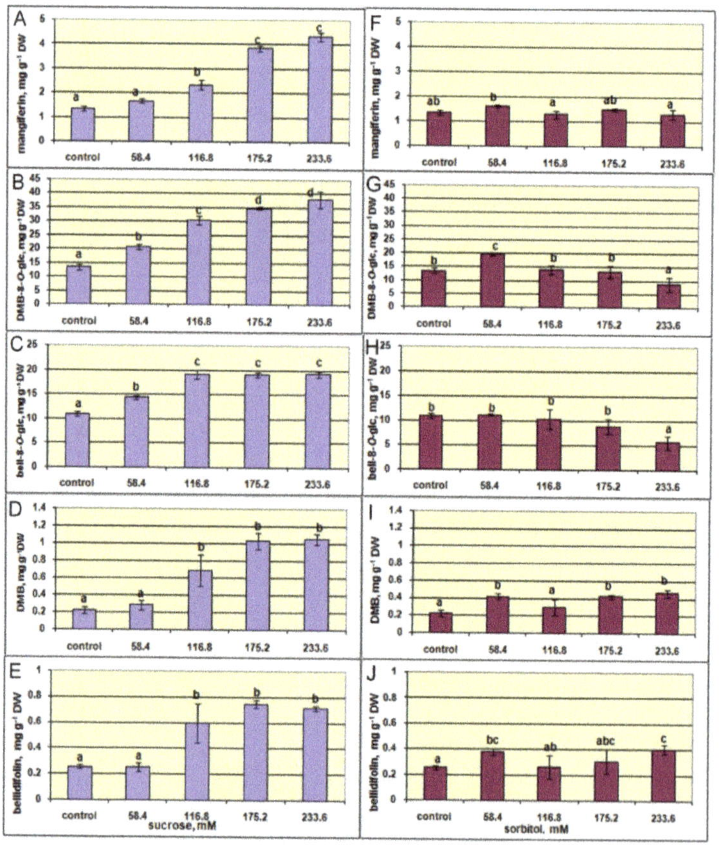

Figure 8. The effect of increasing concentrations (58.4, 161.8, 175.2, and 233.6 mM) of sucrose (**A–E**) and sorbitol (**F–J**) on the content of xanthones mangiferin, demethylbellidifolin-8-*O*-glucoside (DMB-8-*O*-glc), bellidifolin-8-*O*-glucoside (bell-8-*O*-glc), demethylbellidifolin (DMB) and bellidifolin in shoots cultures of *G. lutescens* (line 5) after 35 days of culture. The increasing concentrations of sucrose or sorbitol were added into control medium (BM + 0.2 mg L^{-1} BA containing 58.4 mM sucrose). Values are the means ± SE of four to eleven biological replicates (n = 4–11). Values denoted by the same letter are not significantly different according to the Fisher's LSD test at $p \leq 0.05$ following one-way ANOVA.

The results presented in Figure 8A–E show that xanthone content linearly increased with increasing sucrose concentrations, up to 116.8 mM. With additional increase in sucrose at 175.2 and 233.6 mM, the content of each xanthone reached its highest level. On the other hand, the content of xanthones in shoots cultured with sorbitol was mainly at the level recorded in the control shoots. Compared to the control, only the production of DMB-8-*O*-glc, DMB, and bellidifolin was slightly higher in shoots cultured with the lowest sorbitol concentration (Figure 8G–J).

These results indicate that enhanced accumulation of xanthones in shoots caused by increased concentrations of sucrose might be a consequence of the nutritional effect of sucrose. Numerous studies have shown that higher content of sucrose in the medium stimulated the production of useful secondary metabolites in plant cell cultures. Sucrose is considered to be one of the key sugars in plant life. In addition to its primary building role, sucrose is also an energy supplier for the production of plant biomass. It is also involved in growth, development, storage, signaling, plant stress responses and various metabolic processes [52]. The abundance of sucrose in the culturing medium quite certainly affects and alters metabolic activities in the shoots of *G. lutescens*. The accretion of available source of carbon and energy directs metabolic pathways to formation more complex compounds with a large carbon skeleton, such as xanthones. Further, xanthones may undergo hydroxylation reactions to give xanthone glycosides. As explained in our previous study on the root cultures of *Gentiana dinarica*, higher metabolic activities, due to the increase of available carbohydrates sources as biochemical substrates, may direct xanthone biosynthesis to the enhanced production of xanthone glycosides [28].

The similar positive effect of increased sucrose in the medium on the xanthone content has been reported in *Centaurium erythraea* cultured in vitro [53]. The higher sucrose concentration was more favorable for accumulation of phenols, flavonoids, chlorogenic acid, and total hypericin in adventitious root cultures of *Hypericum perforatum* [54]. The roots of *Gentiana dinarica* produced higher content of xanthones when cultured on medium with more sucrose [28].

2.5. The Effect of Elicitors JA, MeJA and SA on Shoot Growth and Xanthone Production

The elicitation process is indicated as very complex and is subject to many factors, such as elicitor concentration, shoot growth stage, as well as the time of plant tissue exposure to the elicitor [55,56].

In an attempt to enhance xanthones production, JA, MeJA and SA at increasing concentrations (100–500 µM) were applied in shoot culture of *G. lutescens*. Plant tissue responds relatively rapidly (from several hours to several days) to elicitation to induce the synthesis of secondary metabolites [57,58] while prolonged exposure to elicitor may reduce their accumulation over time. For example, the highest production of bacoside in shoots of *Bacopa monnieri* was obtained in the first 7 days of elicitation with MeJA [59] and JA [60]. Guided by the positive results of the elicitation protocol for *Withania somnifera* with SA and MeJA [61], shoot growth and xanthones amounts were analysed in *G. lutescens* shoots cultivated for 7 days on an elicitor-containing medium and another 7 days on an elicitor-free medium.

Figure 9A shows that growth index of the shoots significantly increased after SA low concentration (100 µM) treatment and significantly decreased after treatment with MeJA at the highest one (500 µM), whereas shoot growth was not affected by JA at all concentrations (Figure 9A, Supplementary Material, Table S4). The positive effect of low concentration of SA on *G. lutescens* is not surprising. In general, low concentrations of applied SA promote plant growth under unfavorable conditions. In contrast, high SA concentrations inhibit growth, while the threshold between low and high concentrations depend on plant species. It was shown that SA exhibited growth-promoting (50 µM) and growth-inhibiting (250 µM) effect on *Matricaria chamomilla* seedlings [62].

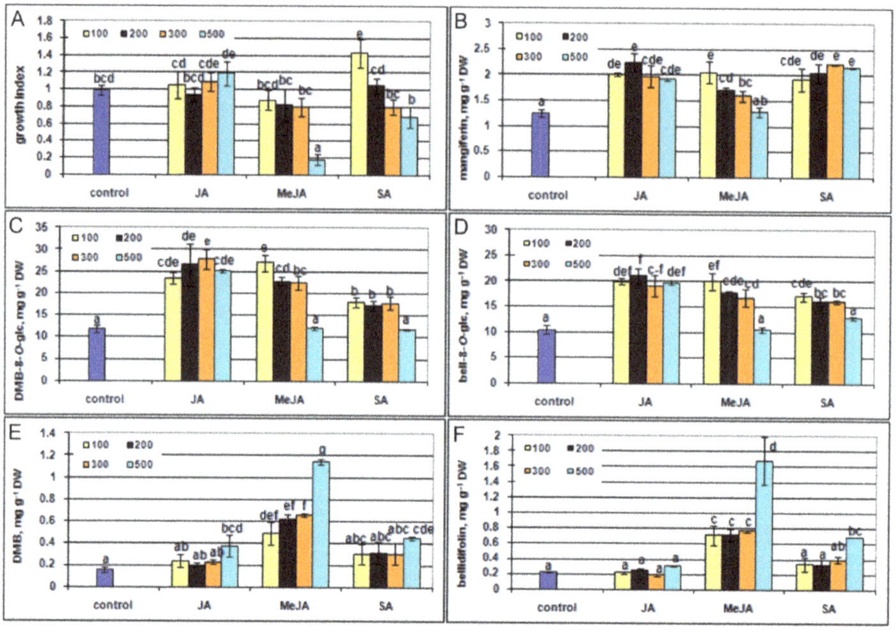

Figure 9. Effect of elicitors jasmonic acid (JA), methyl jasmonat (MeJA) and salicilic acid (SA) at increasing concentrations (100, 200, 300, and 500 µM) on the growth index and xanthone production of shoot cultures of *G. lutescens* (line 5). Growth index (**A**), mangiferin (**B**), DMB-8-*O*-glc (**C**), bell-8-*O*-glc (**D**), DMB (**E**), and bellidifolin (**F**) content. Data represent means ± SE of four to seven biological replicates (n = 4–7). Values denoted by the same letter are not significantly different according to the Fisher's LSD test at $p \leq 0.05$ following ANOVA multifactorial analysis.

SA, JA, and MeJA, were reported to elicit a wide spectrum of phytochemicals in different plant species by inducing the expression of genes for various biosynthetic pathways. Thus, MeJA was reported to stimulate rosmarinic acid accumulation in the cell cultures of *Satureja khuzistannica* [63] and anthocyanins in shoot cultures of *Prunus salicina* × *P. persica* [64], while SA stimulated the production of hypericin and pseudohypericin in the cell cultures of *Hypericum perforatum* [27].

In G. *lutescens* shoot cultures, all tested elicitors contributed almost equally to enhance accumulation of xanthones magniferin, DMB-8-*O*-glc, and bell-8-*O*-glc, while MeJA was superior for elicitation of xanthone aglycons DMB and bellidifolin (Figure 9B–F). The content of xanthone glucosides mangiferin, DMB-8-*O*-glc, and bell-8-*O*-glc at all applied concentration of JA was almost 2-fold higher than in control shoots. However, with the increase in MeJA concentration, production decreased to the level of non-elicited controls. When SA was applied for elicitation, increased concentration did not significantly affect the level of mangiferin, but reduced the content of DMB-8-*O*-glc and bell-8-*O*-glc.

In contrast to xanthone glucosides, the production of aglycones DMB and bellidifolin were not affected or only slightly enhanced upon treatment with lower concentrations of JA and SA. However, shoots elicited with the higher MeJA concentrations showed a significant increase in accumulation of both aglycones. The highest DMB and bellidifolin content (1.14 mg g^{-1} DW and 1.68 mg g^{-1} DW, respectively) was recorded at a concentration of 500 µM MeJA which was 7.4 and 7.6 times higher than in the control shoots, respectively.

A similar positive effect of elicitors on production of xanthone aglycones in hairy roots of *G. dinarica* was reported by Krstić-Milošević et al. [65], where application of biotic elicitors strongly increased production of aglycone norswertianin while simultaneously reducing the production of its glycoside norswertianin-1-*O*-primeveroside. These findings pointed out a selective accumulation of specific xanthone compounds upon treatment with elicitors. In *G. lutescens* shoots they suppressed glycosylation, the late step of the constitutive pathway, leading to increased accumulation of xanthone aglycones. High level of xanthone aglycones is important defensive mechanism for protection the plant cells from oxidative damage, as well as to impair pathogen growth.

Different observations have been recorded regarding the individual efficacy of the particular elicitor compounds. Elicitation effect of MeJA was higher in the production of gymnemic acid in cell suspension of *Gymnema sylvestre* [66], and iridoids and phenylethanoids in the hairy root culture of *Rehmannia glutinosa* [67]. MeJA has also been demonstrated to increase xanthone production in *H. perforatum* cell suspension cultures [25] and in combination with sucrose showed remarkable stimulating effects on hypericin and hyperforin production [68]. On the other hand, higher elicitation effect of SA compared to MeJA was reported for withanolides production from *Withania somnifera* [61]. SA was also more favorable for kinsenoside accumulation and had an equal effect on polysaccharide accumulation with MeJA during *A. roxburghii* rhizome culture [69].

However, many reports suggested that along with the choice of suitable elicitor compound its benefit for the production of metabolite of interest also depended on the concentration used and duration of elicitor treatment.

In *G. lutescens* shoots increased concentration of MeJA showed different effect on accumulation of xanthones. The highest MeJA dose (500 μM) led to the highest DMB and bellidifolin content, while on the other hand strongly reduced the production of xanthone glucosides. A similar effect of SA on accumulation of cardiac glycosides has been shown in shoot cultures of *Digitalis purpurea* [70]. Namely, addition of 50 μM SA promoted production of digitoxin, but increasing concentrations of SA drastically reduced its content. On the contrary, digoxin accumulation was increased in the shoots with an increase in SA concentration up to 200 μM. The flavonoid production of *H. perforatum* cell cultures was significantly promoted by 100 μM MeJA [26], while 4.46 μM MeJA increased the isoflavonoid content in *Pueraria mirifica* cell culture [71]. JA has also been reported as an effective elicitor of secondary metabolite production in many plant species. Thus, enhanced production of oleanolic acid was obtained in cell suspension cultures of *Calendula officinalis* by elicitation with 100 μM JA for 72 h [72], while supplementation of 50 μM JA on day 12 induced the highest anthraquinone content in cell suspension cultures of *Morinda elliptica* [73].

In summary, the findings of the current study suggest that the effectiveness of elicitation of secondary metabolites depends on several factors, including the type of elicitor, the elicitor concentration, the time of exposure to elicitor treatment, and the culture conditions [74].

3. Materials and Methods

3.1. Plant Material

Gentianella lutescens plants at fruitful stage were collected in September 2015 in their native habitat on the southern outskirts of the large mountain field Veliko Košlje at Povlen Mountain, in the locality of Razbojište (latitude 440 10′ 3.43″ north and longitudes 140 37′ 33.474″ east), the Republic of Serbia. A voucher specimen (Co. 6392113/04) was deposited in the Herbarium at the Natural History Museum, Belgrade.

3.2. In Vitro Seed Germination and Shoot Culture Initiation

Immature fruits harvested from collected plants were disinfected in 100 mL 20% (*v*/*v*) commercial bleach solution (4–5% NaOCl) with two drops of liquid detergent (Fairy®) for 25 min and then rinsed 3 times with sterile distilled water.

Immature seeds (1.5 mm) were isolated under stereomicroscope and germinated in Ø90 mm Petri dishes filled with 25 mL of germinating medium consisting of Murashige and Skoog's (MS) mineral salts [75], LS vitamins [76], and 0.64% (*w*/*v*) agar (Institute of Virology, Vaccines and SeraTorlak, Belgrade, Serbia). For shoot culture initiation, the epicotyl explants were isolated from 7-day old seedlings from five seedling lines and placed onto shoot induction medium, which was basal medium (BM) consisting of MS mineral salts, LS vitamins, 100 mg L^{-1} myo-inositol (Sigma-Aldrich, Steinheim, Germany)), 58.4 mM sucrose with addition 0.2 mg L^{-1} BA. The medium was gelled with 0.64% (w/v) agar (Institute of Virology, Vaccines and SeraTorlak, Belgrade, Serbia). Individual epicotyl explant of each seedling line was cultured separately in 100-mL wide-neck Erlenmeyer flask with cotton-wool plugs. Obtained shoots were maintained on the same medium (BM + 0.2 mg L^{-1} BA) and subcultured every 5 weeks on fresh medium. For shoot multiplication, 8–10 shoots of the same line were cultured in Erlenmeyer flask containing basal medium with addition of BA at increasing concentrations 0, 0.05, 0.1, 0.2, 0.5, 1.0, and 2.0 mg L^{-1} for 5 weeks and the number of propagated shoots were recorded. Multiplication index was calculated as main shoot + new axillary shoots (length $\geq$ 2 mm).

For rooting, solid ½ MS medium supplemented with 2% sucrose and increasing (0.2, 0.5, 1.0, 2.0, and 5.0 mg L^{-1}) IBA (indole-3-butyric acid, Sigma-Aldrich) was used. Different durations of root induction treatment were applied, lasting 1–6 days at 5 mg L^{-1} IBA to 50 days at 0.2–2.0 mg L^{-1} IBA (Table 3). For elongation of induced roots, the shoots were transferred into liquid PGR-free ½ MS medium in the test tubes 20 $\times$ 150 mm with metal holders containing $\approx$15 mL of liquid medium. Elongation treatments lasted from 52 to 66 days depending on the concentration of auxin applied (Table 3). Two replicates with 15–26 shoots per rooting treatment were performed and the percentage of rooted shoots and root features were estimated.

All media pH was adjusted to 5.8 prior to autoclaving for 25 min at 114 °C. Cultures were grown at 25 $\pm$ 2 °C in a controlled environment room illuminated with cool-white Phillips fluorescent lamps providing 35–45 µmol m^{-2} s^{-1} under a 16-h (long day) photoperiod.

3.3. Increase in Sucrose and Sorbitol Level

To examine the effect of enhanced osmotic pressure on the shoot growth and xanthone production, 400 mg of *G. lutescens* shoots were cultivated per Erlenmeyer flask on BM + 0.2 mg L^{-1} BA wherein sucrose or sorbitol at increasing concentrations 58.4, 116.8, 175.2, 233.6 mM were added. Shoot growth was measured after 5 weeks of culture in terms of fresh weight and dry weight of shoots per Erlenmeyer flask. Growth index [(final fresh weight − initial fresh weight)/initial fresh weight], % dry matter [(dry weight/final fresh weight) $\times$ 100], and accumulation of xanthones were determined for the harvested shoots.

3.4. Elicitor Preparation and Application

Jasmonic acid (JA, Duchefa), methyl jasmonate (MeJA, Duchefa) and salicylic acid (SA, Duchefa) 10 mM stock solutions were prepared in 50% (v/v) ethanol and then filter-sterilized using 0.22 µm filter. The shoots (400 mg weight) were isolated from 35–40-day-old cultures and grown in BM + 0.2 mg L^{-1} BA supplemented with elicitors at the following final concentrations: 100, 200, 300, and 500 µM. The treatment with elicitor lasted 7 days, and then the shoots were cultured on a control medium for another 7 days. Control shoots were cultured on BM + 0.2 mg L^{-1} BA without elicitors. At the end of the experiment growth index [(final fresh weight − initial fresh weight)/initial fresh weight] and accumulation of xanthones were determined for the harvested shoots.

3.5. Xanthone Extraction and Conditions for HPLC-DAD Analysis

Xanthone compounds extraction from dry powered plant material was performed as previously reported by Krstić-Milošević et al. [77]. Obtained extracts were filtered into volumetric flasks (10 mL) and adjusted to the volume with methanol. Prior to HPLC analysis, extracts were filtered through a nylon syringe filters (Captiva syringe filters, 0.45 mm, 13 mm, Agilent Technologies, Waldbronn, Germany). Chromatographic analysis was carried out on Agilent series 1100 HPLC instrument (Agilent Technologies, Waldbronn, Germany), with a diode array detector, on a reverse phase Zorbax SB-C18 (Agilent Technologies, Waldbronn, Germany) analytical column (150 mm $\times$ 4.6 mm i.d., 5 µm particle size) thermostated at 25 °C. The mobile phase consisted of two solvents: A (1% v/v solution of orthophosphoric acid in water) and B (acetonitrile) in the following gradient elution: 98–90% A 0–5 min, 90% A 5–13 min, 90–75% A 13–15 min, 75% A 15–18 min, 75–70% A 18–20 min, 70–40% A 20–24 min, 40–0% A 24–28 min. The injection volume was 5 µL. Detection wavelengths were set at 260 and 320 nm, and the flow rate was 1 mL min^{-1}. The isolation, identification, and characterization of xanthones demethylbellidifolin-8-*O*-glucoside (DMB-8-*O*-glc), bellidifolin-8-*O*-glucoside (bell-8-*O*-glc), demethylbellidifolin (DMB) and bellidifolin were reported in previous study [40]. Mangiferin was purchased from Sigma-Aldrich (Steinheim, Germany). Quantification was performed using standardized calibration curves of xanthones. The content of xanthones in the samples was determined by calculation of peak area and expressed as milligrams per gram of dry weight.

3.6. Statistical Analysis

All in vitro culture experiments were repeated at least 2–4 times with 15–40 shoots/explants per treatment. Biochemical analyses were repeated 4–11 times. The data were subjected to standard one-way analysis of variance (ANOVA) except the data related to the effects of elicitors where two-way ANOVA was applied. Percentage data were subjected to angular transformation before statistical analysis, followed by inverse transformation for presentation. Means and standard errors were calculated for numerical parameters and their differences was analyzed by *t*-test or Fisher's multiple range LSD test at $p \leq 0.05$ using the StatGraphics Plus software package for Windows 2.1 (Statistical Graphics Corp., Rockville, MD, USA).

4. Conclusions

The present study represents the first achievement in establishing an in vitro culture of *G. lutescens* that can be used for mass shoot propagation, significantly contributing to the preservation of this valuable medicinal plant. This is also the first report on the composition of pharmacologically active xanthones in wild and tissue cultured shoots of *G. lutescens.* The promising in vitro protocol included sterile germination of immature seeds on PGR-free MS medium, establishment of shoot culture from epicotyl explants isolated from seedlings on MS + 0.2 mg L^{-1} BA, as well as shoot multiplication and elongation on the same substrate. Satisfactory rooting included two phases, root induction on ½ MS medium with 5.0 mg L^{-1} IBA for 2 days, followed by root elongation in liquid PGR-free ½ MS medium for 60 days. The results indicated that the use of increased sucrose, as well as abiotic elicitors JA, MeJA and SA stimulated the accumulation of bioactive xanthones in tissue-cultured shoots. Multiple increases of the amount of aglycones BMD and bellidifolin was achieved by applying MeJa at the highest concentration. Since in vitro shoot cultures of *G. lutescens* were proved as a prospective modality for the accumulation of xanthones of significant interest, advanced conditions will be applied in further studies for the production of metabolites in large-scale in bioreactors. These achievements increase the commercial prospects of *G. lutescens* in the medicinal plant industry.

Supplementary Materials: The following are available online at https://www.mdpi.com/article/10.3390/plants10081651/s1, Figure S1: The effect of increasing concentrations of BA (0–2 mg L^{-1}) on the content of xanthones mangiferin (A), demethylbellidifolin-8-*O*-glucoside (DMB-8-*O*-glc) (B), bellidifolin-8-*O*-glucoside (bell-8-*O*-glc) (C), demethylbellidifolin (DMB) (D), and bellidifolin (E) in shoots cultures of *G. lutescens* line 5 after 35 days of cultivation. Values are the means ± SE of six biological replicates (n = 6). Data were analysed by one-way ANOVA (F). Values followed by different letters are significantly different according to Fisher's LSD test at $p \leq 0.05$. Table S1: The results of nested ANOVA for the effects of increasing sucrose and sorbitol concentrations on the growth parameters (growth index, fresh and dry weight and % of dry matter) of shoot cultures of *G. lutescens* line 5. The bold values indicate statistically significant differences ($p \leq 0.05$). Table S2: The results of two-way ANOVA for the influence of carbohydrate type (sucrose or sorbitol) and their concentrations on in vitro growth and flowering of shoots of *G. lutescens*. The bold values indicate statistically significant differences ($p \leq 0.05$). Table S3: The results of nested ANOVA for the effects of increasing sucrose and sorbitol concentrations on the xanthones production in shoot cultures of

G. lutescens line 5. The bold values indicate statistically significant differences ($p \leq 0.05$). Table S4: The results of nested ANOVA for the effects of elicitor type and concentrations on the shoot growth and xanthones production in shoot cultures of *G. lutescens* line 5. The bold values indicate statistically significant differences ($p \leq 0.05$).

Author Contributions: Conceptualization, B.V. and D.K.-M.; methodology, B.V., D.V. and D.K.-M.; validation, N.B. and D.K.-M.; formal analysis, N.B.; investigation, T.J., D.K.-M. and B.V.; data curation, B.V. and N.B.; writing—original draft preparation, N.B., B.V. and D.K.-M.; visualization, T.J.; supervision, B.V. All authors have read and agreed to the published version of the manuscript.

Funding: This research was funded by the Ministry of Education, Science and Technological Development, Republic of Serbia under contract 451-03-9/2021-14/200007.

Institutional Review Board Statement: Not applicable.

Informed Consent Statement: Not applicable.

Data Availability Statement: The data presented in this study are available from the authors.

Conflicts of Interest: The authors declare no conflict of interest.

References

1. von Hagen, K.B.; Kadereit, J.W. The phylogeny of *Gentianella* (Gentianaceae) and its colonization of the southern hemisphere as revealed by nuclear and chloroplast DNA sequence variation. *Org. Divers. Evol.* **2001**, *1*, 61–79. [CrossRef]
2. Stevanović, V.; Jakovljević, K. Distribution, ecology, and some taxonomical notes of the genera *Gentiana* L. and *Gentianella* Moench (Gentianaceae) in the Balkans. In *The Gentianaceae—Volume 1: Characterization and Ecology*; Rybczyński, J., Davey, M., Mikula, A., Eds.; Springer: Berlin/Heidelberg, Germany, 2014; pp. 169–200.
3. Jovanović-Dunjić, R. Rod *Gentianella* Moench. In *Flora SR Srbije*; Josifović, M., Ed.; SANU: Beograd, Serbia, 1973; Volume 5, pp. 426–433.
4. Lacaille-Dubois, M.; Galle, K.; Wagner, H. Secoiridoids and xanthones from *Gentianella nitida*. *Planta Med.* **1996**, *62*, 365–367. [CrossRef]
5. Nadinic, E.; Gorzalczany, S.; Rojo, A.; van Baren, C.; Debenedetti, S.; Acevedo, C. Topical antiinflammatory activity of *Gentianella achalensis*. *Fitoterapia* **1999**, *70*, 166–171. [CrossRef]
6. Lu, S.; Tanaka, N.; Kawazoe, K.; Murakami, K.; Damdinjav, D.; Dorjbal, E.; Kashiwada, Y. Tetrahydroxanthones from Mongolian medicinal plant *Gentianella amarella* ssp. *acuta*. *J. Natur. Med.* **2016**, *70*, 780–788. [CrossRef]
7. Hostettmann-Kaldas, M.; Jacot-Guillarmod, A. Contribution to the phytochemistry of the genus *Gentiana*. Part XXIII. Xanthones and flavone C-glucosides of the genus *Gentiana* (subgenus *Gentianella*). *Phytochemistry* **1978**, *17*, 2083–2086. [CrossRef]
8. Tovilovic-Kovacevic, G.; Zogovic, N.; Krstic-Milosevic, D. Secondary metabolites from endangered *Gentiana, Gentianella, Centaurium,* and *Swertia* species (Gentianaceae): Promising natural biotherapeutics. In *Biodiversity and Biomedicine*; Ozturk, M., Egamberdieva, D., Pešić, M., Eds.; Elsevier-Academic Press: London, UK, 2020; pp. 335–385.
9. Tian, L.-Y.; Bai, X.; Chen, X.-H.; Fang, J.-B.; Liu, S.-H.; Chen, J.-C. Anti-diabetic effect of methylswertianin and bellidifolin from *Swertia punicea* Hemsl. and its potential mechanism. *Phytomedicine* **2010**, *17*, 533–539. [CrossRef]
10. Zhao, Z.-Y.; Gao, Y.-Y.; Gao, L.; Zhang, M.; Wang, H.; Zhang, C.-H. Protective effects of bellidifolin in hypoxia-induced in pheochromocytoma cells (PC12) and underlying mechanisms. *J. Toxicol. Envir. Health Part A* **2017**, *80*, 1187–1192. [CrossRef]
11. Cruz, M.I.; Cidade, H.; Pinto, M. Dual/multitargeted xanthone derivatives for Alzheimer's disease: Where do we stand? *Future Med. Chem.* **2017**, *9*, 1611–1630. [CrossRef] [PubMed]
12. Tutin, T.G. Gentianella Moench. In *Flora Europea*; Tutin, T.G., Heywood, V.H., Burges, N.A., Moore, D.M., Valentine, D.H., Walters, S.M., Webb, D.A., Eds.; Cambridge University Press: Cambridge, UK, 1972; Volume 3, pp. 63–67.
13. Kirschnerová, L.; Kirschner, J. Changes in the distribution of *Gentianella lutescens* subsp. *lutescens* and subsp. *carpatica* in the Czech Republic. *Zprávy Èes. Bot. Spoleè. Praha* **2003**, *38*, 205–216.
14. IUCN. *IUCN Red List Categories and Criteria: Version 3.1*, 2nd ed.; IUCN: Gland, Switzerland; Cambridge, UK, 2012; 32p.
15. Sharma, V.; Bhardwaj, U.; Sharma, S.; Sharma, S. Medicinal Plants: Need for Sustainable Exploitation (With Special reference to Himachal Pradesh). *J. Pharm. Res.* **2012**, *5*, 4313–4317.
16. Dias, M.I.; Sousa, M.J.; Alves, R.C.; Ferreira, I.C.F.R. Exploring plant tissue culture to improve the production of phenolic compounds: A review. *Ind. Crops Prod.* **2016**, *82*, 9–22. [CrossRef]
17. Chandran, H.; Meena, M.; Barupal, T.; Sharma, K. Plant tissue culture as a perpetual source for production of industrially important bioactive compounds. *Biotechn. Rep.* **2020**, *26*, e00450. [CrossRef]
18. Vinterhalter, B.; Janković, T.; Šavikin, K.; Nikolić, R.; Vinterhalter, D. Propagation and xanthone content of *Gentianella austriaca* shoot cultures. *Plant. Cell Tiss. Organ. Cult.* **2008**, *94*, 329–335. [CrossRef]
19. Janković, T.; Vinterhalter, B.; Krstić-Milošević, D.; Nikolić, R.; Vinterhalter, D.; Milosavljević, S. Xanthone compounds in shoot cultures of *Gentianella bulgarica*. *Acta Physiol. Plant.* **2011**, *33*, 1515–1520. [CrossRef]

20. Krstić-Milošević, D.; Vinterhalter, B.; Janković, T.; Vinterhalter, D. Biotechnology and Phytochemistry of *Gentianella* Species from the Central Regions of the Balkan Peninsula. In *The Gentianaceae—Biotechnology and Applications*; Rybczyński, J., Davey, M., Mikuła, A., Eds.; Springer: Berlin/Heidelberg, Germany, 2015; Volume 2, pp. 93–112.
21. Huo, L.-Y.; Zheng, X.-J. Studies on tissue culture of *Gentianella albifpra*. *Chin. Pharm. J.* **2002**, *37*, 415–418.
22. Solorzano Acosta, A.; Ruiz Reyez, G.; Venegas, E. Preliminary identification secondry metabolites produced in induced by callus "Corpus way" *Gentianella bicolor* (Wedd.) Fabris exJ.s. Pringle. *Rev. Sci.* **2014**, *6*, 137–142.
23. Moraes, R.M.; Cerdeira, A.L.; Lourenço, M.V. Using micropropagation to develop medicinal plants into crops. *Molecules* **2021**, *26*, 1752. [CrossRef] [PubMed]
24. Franklin, G.; Conceicao, L.F.R.; Kombrink, E.; Dias, A.C.P. Xanthone biosynthesis in *Hypericum perforatum* cells provides antioxidant and antimicrobial protection upon biotic stress. *Phytochemistry* **2009**, *70*, 65–73. [CrossRef]
25. Conceição, L.F.R.; Ferreres, F.; Tavares, R.M.; Dias, A.C.P. Induction of *Hypericum perforatum* L. cells by *Colletotrichum gloeosporioides* elicitation. *Phytochemistry* **2006**, *67*, 140–155. [CrossRef]
26. Wang, J.; Qian, J.; Yao, L.; Lu, Y. Enhanced production of flavonoids by methyl jasmonate elicitation in cell suspension culture of *Hypericum perforatum*. *Bioresour. Bioprocess.* **2015**, *2*, 5–13. [CrossRef]
27. Gadzovska, S.; Maury, S.; Delaunay, A.; Spasenoski, M.; Hagege, D.; Courtois, D.; Joseph, C. The influence of salicylic acid elicitation of shoots, callus, and cell suspension cultures on production of naphtodianthrones and phenylpropanoids in *Hypericum perforatum* L. *Plant. Cell Tiss. Org. Cult.* **2013**, *113*, 25–39. [CrossRef]
28. Krstić-Milošević, D.; Janković, T.; Vinterhalter, B.; Menković, N.; Aljančić, I.; Vinterhalter, D. Influence of carbohydrate source on xanthone content in root cultures of *Gentiana dinarica* Beck. *Plant. Growth Regul.* **2013**, *71*, 147–155. [CrossRef]
29. Sahitya, U.L.; Krishna, M.S.R.; Deepthi, R.S. Biochemical and physiological changes induced by water stress in hot pepper (*Capsicum annum* L.) genotypes. *Plant. Cell Biotech. Mol. Biol.* **2018**, *19*, 179–195.
30. Jensen, S.R.; Schripsema, J. Chemotaxonomy and pharmacology of Gentianaceae. In *Gentianaceae: Systematics and Natural History*; Struwe, L., Albert, V.A., Eds.; Cambridge University Press: Cambridge, UK, 2002; p. 573.
31. Janković, T.; Krstić, D.; Aljančić, I.; Šavikin-Fodulović, K.; Menković, N.; Vajs, V.; Milosavljević, S. Xanthones and C-glucosides from the aerial parts of four species of *Gentianella* from Serbia and Montenegro. *Biol. Biochem. Syst. Ecol.* **2005**, *33*, 729–735. [CrossRef]
32. Murthy, H.N.; Dandin, V.S.; Paek, K.-Y. Tools for biotechnological production of useful phytochemicals from adventitious root cultures. *Phytochem. Rev.* **2016**, *15*, 129–145. [CrossRef]
33. Bakhtiar, Z.; Mirjalili, M.H.; Sonboli, A.; Moridi Farimani, M.; Ayyari, M. In vitro propagation, genetic and phytochemical assessment of *Thymus persicus*—A medicinally important source of pentacyclic triterpenoids. *Biologia* **2014**, *69*, 594–603. [CrossRef]
34. Sirikonda, A.; Jogam, P.; Ellendula, R.; Kudikala, H.; Mood, K.; Allini, V.R. In vitro micropropagation and genetic fidelity assesment in *Flemingia macrophylla* (Willd.) Merr: An ethnomedicinal plant. *Vegetos* **2020**, *33*, 286–295. [CrossRef]
35. Devi, T.R.; Dasgupta, M.; Sahoo, M.R.; Kole, P.C.; Prakash, N. High efficient *de novo* rootto-shoot organogenesis in *Citrus jambhiri* Lush.: Gene expression, genetic stability and virus indexing. *PLoS ONE* **2021**, *16*, e0246971. [CrossRef] [PubMed]
36. Gu, R.H.; Hong, L.Y.; Long, C.L. The ways of producing secondary metabolites via plant cell culture. *Plant. Physiol. J.* **2016**, *49*, 869–881.
37. Dević, M.; Momčilović, I.; Kristić, D.; Maksimović, V.; Konjević, R. In vitro multiplication of willow gentian (*Gentiana asclepiadea* L.) and the production of gentiopicrin and mangiferin. *Phyton* **2006**, *46*, 45–54.
38. Rolland, F.; Moore, B.; Sheen, J. Sugar sensing and signaling in plants. *Plant. Cell* **2002**, *14*, 185–205. [CrossRef]
39. De Paiva Neto, V.B.; Otoni, W.C. Carbon sources and their osmotic potential in plant tissue culture: Does it matter? *Sci. Hortic.* **2003**, *97*, 193–202. [CrossRef]
40. Grattapaglia, D.; Machado, M.A. Micropropagação. In *Cultura de Tecidos e Transformação Genética de Plantas*; Torres, A.C., Caldas, L.S., Buso, J.A., Eds.; Embrapa-SPI/Embrapa-CNPH: Brasília, Brazil, 1998; Volume 1, pp. 43–76.
41. Hazubska-Przybył, T.; Kalemba, E.; Ratajczak, E.; Bojarczuk, K. Effects of abscisic acid and an osmoticum on the maturation, starch accumulation and germination of *Picea* spp. somatic embryos. *Acta Phy. Plant.* **2016**, *38*, 59. [CrossRef]
42. Siwach, P.; Grover, K.; Gill, A.R. The influence of plant growth regulators, explant nature and sucrose concentration on in vitro callus growth of *Thevetia peruviana* Schum. *Asian J. Biotechnol.* **2011**, *3*, 280–292. [CrossRef]
43. Takeno, K. Stress-induced flowering: The third category of flowering response. *J. Exp. Bot.* **2016**, *67*, 4925–4934. [CrossRef]
44. Al-Khateeb, A.A. Regulation of in vitro bud formation of date palm (*Phoenix dactylifera* L.) cv. Khanezi by different carbon sources. *Bioresour. Technol.* **2008**, *99*, 6550–6555. [CrossRef]
45. Mohamed, M.A.H.; Alsadon, A.A. Influence of ventilation and sucrose on growth and leaf anatomy of micropropagated potato plantlets. *Sci. Hort.* **2010**, *123*, 295–300. [CrossRef]
46. Martins, J.P.R.; Verdoodt, V.; Pasqual, M.; De Proft, M. Impacts of photoautotrophic and photomixotrophic conditions on in vitro propagated *Billbergia zebrina* (Bromeliaceae). *Plant. Cell Tiss. Org. Cult.* **2015**, *123*, 121–132. [CrossRef]
47. Sivanesan, L.; Hwang, S.L.; Jeong, B.R. Influence of plant growth regulators on axillary shoot multiplication and iron source on growth of *Scrophularia takesimensis* Nakal—A rare endemic medicinal plant. *Afr. J. Biotechnol.* **2008**, *7*, 4484–4490.
48. Babb, V.M.; Haigler, C.H. Sucrose phosphate synthase activity rises in correlation with high-rate cellulose synthesis in three heterotrophic systems. *Plant. Physiol.* **2001**, *127*, 1234–1242. [CrossRef]

49. Gurel, S.; Gulsen, Y. The effects of different sucrose, agar and pH levels on in vitro shoot production of almond (*Amygdalus communis* L.). *Turkish J. Bot.* **1998**, *22*, 363–373.
50. Sotiropoulos, T.E.; Molassiotis, A.N.; Mouhtaridou, G.I.; Papadakis, I.; Dimassi, K.N.; Therios, I.N.; Diamantidis, G. Sucrose and sorbitol effects on shoot growth and proliferation in vitro, nutritional status and peroxidase and catalaze isoenzymes of M 9 and MM 106 apple (*Malus domestica* Borkh.) rootstocks. *Eur. J. Hortic. Sci.* **2006**, *71*, 114–119.
51. Ahmad, T.; Abbasi, N.A.; Hafiz, I.A.; Ali, A. Comparison of sucrose and sorbitol as main carbon energy sources in micropropagation of peach rootstock GF-677. *Pak. J. Bot.* **2007**, *39*, 1269–1275.
52. Salerno, G.L.; Curatti, L. Origin of sucrose metabolism in higher plants: When, how and why? *Trends Plant. Sci.* **2003**, *8*, 63–69. [CrossRef]
53. Janković, T.; Krstić, D.; Šavikin-Fodulović, K.; Menković, N.; Grubišić, D. Xanthone compounds of *Centaurium erythraea* grown in nature and cultured in vitro. *Pharm. Pharmacol. Lett.* **2000**, *10*, 23–25.
54. Cui, X.-H.; Murthy, H.N.; Wu, C.-H.; Peak, K.-Y. Sucrose-induced osmotic stress affects biomass, metabolite, and antioxidant levels in root suspension cultures of *Hypericum perforatum* L. *Plant. Cell Tiss. Org. Cult.* **2010**, *103*, 7–14. [CrossRef]
55. Zhao, J.; Davis, L.C.; Verpoorte, R. Elicitor signal transduction leading to production of plant secondary metabolites. *Biotechnol. Adv.* **2005**, *23*, 283–333. [CrossRef]
56. Namedo, A.G. Plant cell elicitation for production of secondary metabolites: A review. *Pharmacogn. Rev.* **2007**, *1*, 69–79.
57. Wungsintaweekul, J.; Choo-malee, J.; Charoonratana, T.; Niwat, N. Methyl jasmonate and yeast extract stimulate mitragynine production in *Mitragyna speciosa* (Roxb.) Korth shoot culture. *Biotechnol. Lett.* **2012**, *34*, 1945–1950. [CrossRef]
58. Ziaratnia, S.; Kunert, K.; Lall, N. Elicitation of 7-methyljuglone in *Drosera capensis*. *S. Afr. J. Bot.* **2009**, *75*, 97–103. [CrossRef]
59. Sharma, P.; Yadav, S.; Srivastava, A.; Shrivastava, N. Methyl jasmonate mediates upregulation of bacoside A production in shoot cultures of *Bacopa monnieri*. *Biotechnol. Lett.* **2013**, *35*, 1121–1125. [CrossRef]
60. Munish, S.; Ashok, A.; Rajinder, G.; Sharada, M. Enhanced bacoside production in shoot cultures of *Bacopa monnieri* under the influence of abiotic elicitors. *Nat. Prod. Res.* **2015**, *29*, 745–749.
61. Sivanandhan, G.; Rajesh, M.; Arun, M.; Jeyaraj, M.; Kapil Dev, G.; Arjunan, A.; Manickavasagam, M.; Muthuselvam, M.; Selvaraj, N.; Ganapathi, A. Effect of culture conditions, cytokinins, methyl jasmonate and salicylic acid on the biomass accumulation and production of withanolides in multiple shoot culture of *Withania somnifera* (L.) Dunal using liquid culture. *Acta Physiol. Plant.* **2013**, *35*, 715–728. [CrossRef]
62. Kovácik, J.; Grúz, J.; Backor, M.; Strnad, M.; Repcák, M. Salicylic acid induced changes to growth and phenolic metabolism in *Matricaria chamomilla* plants. *Plant. Cell Rep.* **2009**, *28*, 135–143. [CrossRef]
63. Khojasteh, A.; Mirjalili, M.H.; Palazon, J.; Eibl, R.; Cusido, R.M. Methyl jasmonate enhanced production of rosmarinic acid in cell cultures of *Satureja khuzistanica* in a bioreactor. *Eng. Life Sci.* **2016**, *16*, 740–749. [CrossRef]
64. Lucioli, S.; Di Bari, C.; Nota, P.; Frattarelli, A.; Forni, C.; Caboni, E. Methyl jasmonate promotes anthocyanins' production in *Prunus salicina* × *Prunus persica* in vitro shoot cultures. *Plant. Biosyst.* **2017**, *151*, 788–791. [CrossRef]
65. Krstić-Milošević, D.; Janković, T.; Uzelac, B.; Vinterhalter, D.; Vinterhalter, B. Effect of elicitors on xanthone accumulation and biomass production in hairy root cultures of *Gentiana dinarica*. *Plant. Cell Tiss. Org. Cult.* **2017**, *130*, 631–664. [CrossRef]
66. Chodisetti, B.; Rao, K.; Gandi, S.; Giri, A. Gymnemic acid enhancement in the suspension cultures of *Gymnema sylvestre* by using the signaling molecules-methyl jasmonate and salicylic acid. *In Vitro Cell Dev. Biol. Plant* **2015**, *51*, 88–92. [CrossRef]
67. Piatczak, E.; Kuzma, L.; Wysokinska, H. The influence of methyl jasmonate and salicylic acid on secondary metabolite production in *Rehmannia glutinosa* libosch hairy root culture. *Acta Biol. Cracov. Bot.* **2016**, *58*, 57–65. [CrossRef]
68. Pavlik, M.; Vacek, J.; Klejdus, B.; Kuban, V. Hypericin and hyperforin production in St. John's worth in vitro culture: Influence of saccharose, polyethylene glycol, methyl jasmonate and *Agrobacterium tumefaciens*. *J. Agric. Food Chem.* **2007**, *55*, 6147–6153. [CrossRef] [PubMed]
69. Luo, W.-Y.; Yang, F.; Piao, X.-C.; Jin, M.-Y.; Tian, W.; Gao, Y.; Lian, M.-L. Promising strategy to efficiently improve the kinsenoside and polysaccharide production of rhizome cultures of *Anoectochilus roxburghii* (Wall.) Lindl. *Ind. Crops Prod.* **2018**, *125*, 269–275. [CrossRef]
70. Patil, J.G.; Ahire, M.L.; Nitnaware, K.M.; Panda, S.; Bhatt, V.P.; Kavi Kishor, P.B.; Nikam, T.D. In vitro propagation and production of cardiotonic glycosides in shoot cultures of *Digitalis purpurea* L. by elicitation and precursor feeding. *Appl. Microbiol. Biotechnol.* **2013**, *97*, 2379–2393. [CrossRef]
71. Tanatorn, S.; Nuttha, T.; Polkit, S.; Aphichart, K. Effect of methyl jasmonate on isoflavonoid accumulation and antioxidant enzymes in *Pueraria mirifica* cell suspension culture. *J. Med. Plant. Res.* **2014**, *8*, 401–407. [CrossRef]
72. Wiktorowska, E.; Dlugosz, M.; Janiszowska, W. Significant enhancement of oleanolic acid accumulation by biotic elicitors in cell suspension cultures of *Calendula officinalis* L. *Enzym Microb. Technol.* **2010**, *46*, 14–20. [CrossRef]
73. Chong, T.M.; Abdullah, M.A.; Lai, O.M.; Nor'Aini, F.M.; Lajis, N.H. Effective elicitation factors in *Morinda elliptica* cell suspension culture. *Process. Biochem.* **2005**, *40*, 3397–3405. [CrossRef]
74. Coste, A.; Vlase, L.; Halmagyi, A.; Deliu, C.; Coldea, G. Effects of plant growth regulators and elicitors on production of secondary metabolites in shoot cultures of *Hypericum hirsutum* and *Hypericum maculatum*. *Plant. Cell Tiss. Org. Cult.* **2011**, *106*, 279–288. [CrossRef]
75. Murashige, T.; Skoog, F.A. Revised medium for rapid growth and bioassays with tobacco tissue cultures. *Physiol. Plant.* **1962**, *15*, 473–497. [CrossRef]

76. Linsmaier, E.M.; Skoog, F. Organic growth factor requirements of tobacco tissue cultures. *Physiol. Plant.* **1965**, *18*, 100–127. [CrossRef]
77. Krstić-Milošević, D.; Banjac, N.; Janković, T.; Eler, K.; Vinterhalter, B. *Gentiana clusii* Perr.&Song.: Enhanced production of secondary metabolites by in vitro propagation. *Plant. Phyisol. Biochem.* **2020**, *154*, 735–744.

plants

MDPI

Article

In Vitro Propagation Method for Production of Phenolic-Rich Planting Material of Culinary Rhubarb 'Malinowy'

Agnieszka Wojtania [1,*] and Monika Mieszczakowska-Frąc [2]

[1] Department of Applied Biology, The National Institute of Horticultural Research, Konstytucji 3 Maja 1/3 Str., 96-100 Skierniewice, Poland

[2] Fruit and Vegetable Storage and Processing Department, The National Institute of Horticultural Research, Konstytucji 3 Maja 1/3 Str., 96-100 Skierniewice, Poland; monika.mieszczakowska@inhort.pl

* Correspondence: agnieszka.wojtania@inhort.pl

Abstract: Culinary rhubarb is a popular vegetable crop, valued for its long, thickened stalks, very rich in different natural bioactive ingredients. Tissue cultures are a useful tool for vegetative propagation of virus-free rhubarb plants and rapid multiplication of valuable selected genotypes. The aim of this study was to develop an effective method for in vitro propagation of selected genotypes of Polish rhubarb 'Malinowy' characterized by high yield and straight, thick and intensive red stalks. Identification and quantification of anthocyanins and soluble sugars by the HPLC method in shoot cultures and ex vitro established plantlets were also performed. Shoot cultures were established from axillary buds isolated from dormant, eight-year-old rhizomes. Effective shoot multiplication of rhubarb 'Malinowy' was obtained in the presence of 6.6 μM benzylaminopurine or 12.4 μM *meta*-topolin. Both cytokinins stimulated shoot formation in a manner that depended on sucrose concentration. Increasing the sucrose concentration from 59 to 175 mM decreased the production of shoots and outgrowth of leaves by 3-fold but enhanced shoot length, single shoot mass and callus formation at the base of shoots. This coincided with increased accumulation of soluble sugars (fructose, glucose) and anthocyanins-cyanidin-3-O-rutinoside (max. 208.2 mg·100 g^{-1} DM) and cyanidin-3-O-glucoside (max. 47.7 mg·100 g^{-1} DM). The highest rooting frequency (94.9%) and further successful ex vitro establishment (100%) were observed for shoots that were earlier rooted in vitro in the presence of 4.9 μM indole-3-butyric acid. Our results indicated that anthocyanin contents in leaf petioles were influenced by developmental stage. Under in vitro conditions, it is possible to elicit those pigments by sucrose at high concentration and *meta*-topolin.

Keywords: anthocyanins; *meta*-topolin; micropropagation; *Rheum*; soluble sugars; sucrose concentration

Citation: Wojtania, A.; Mieszczakowska-Frąc, M. In Vitro Propagation Method for Production of Phenolic-Rich Planting Material of Culinary Rhubarb 'Malinowy'. *Plants* **2021**, *10*, 1768. https://doi.org/10.3390/plants10091768

Academic Editors: Laura Pistelli, Kalina Danova and Iyyakkannu Sivanesan

Received: 23 June 2021
Accepted: 20 August 2021
Published: 25 August 2021

1. Introduction

Rhubarb (*Rheum*) is a herbaceous perennial of the *Polygonaceae* family. The genus *Rheum* includes about 60 species, and most of them are native to the northern and central regions of Asia [1]. For thousands of years, *Rheum* has been cultivated in China for medicinal purposes. The dried rhizomes and roots of medicinal species (*R. palmatum* L., *R. officinale* Baill, *R. tanguticum* Maxim.) have been used to treat constipation, inflammation and ulcers [2,3].

Culinary rhubarb is popular as a vegetable crop, valued for its long, thickened stalks [4]. They are used in the production of desserts, cakes, jam, juices, fruit teas and wine. The use of rhubarb petioles for food was discovered at the beginning of the 18th century in Great Britain, but widespread consumption of rhubarb stalks began in the early 19th century. Culinary rhubarb cultivation spread to northern Europe, North America, Australia and New Zealand. Worldwide rhubarb production and consumption peaked just before the Second World War, then it came to be restricted [5]. Recently, there has been an increased interest in rhubarb production in Europe, including in Poland. This is due to the higher interest of consumers and the food industry in functional food [6]. It is known that rhubarb stalks contain useful levels of organic acids (malic acid, citric acid, fumaric acid and ascorbic acid), dietary fiber, protein, potassium, calcium and magnesium [7]. They are also rich in polyphenolic compounds, such as stilbenes, anthocyanins and flavonols, which have a range of bioactivities relevant to human health [8–10]. Rhubarb juice may be used as a natural food preservative. It has been shown that the addition of rhubarb juice to strawberry jams or apple purees reduced unfavorable color changes and increased antioxidant properties [7]. Consumers and the food industry prefer red stalks that are sweeter and have a higher content of polyphenolics. It has been shown that the level and composition of phenolic compounds depend on the rhubarb cultivar [11].

The culinary rhubarb called *R. rhaponticum* L. is a hybrid between rhubarb species originally brought to Europe for medicinal purposes—*R. rhaponticum*, *R. undulatum* L. (syn. *R. rhabarbarum* L.) and *R. palmatum* L. [5,12]. At an early stage, growers in England started to develop culinary rhubarb cultivars, including the still-common 'Victoria' and 'Prince Albert'. Nowadays, the number of culinary rhubarb cultivars is difficult to estimate. Propagation of culinary cultivars by seed and careful selection of superior plants resulted in the development of many local cultivars with marked improvement in yield, quality and uniformity.

For successful commercial cultivation of rhubarb genotypes, a planting material of good quality is very important. In horticultural practice, seed propagation is not recommended because the obtained plants are usually not uniform and do not repeat parental traits. Plants differ in height, habit, shape and intensity of petiole color, as well as petiole size and number [13]. On the other hand, the production of rhubarb by division of crowns is limited by the low number of donor plants available and the risk of virus transfer. Rhubarb crops were found to be infected with several viruses. The most common are: *Turnip mosaic virus* (TuMV), *Arabis mosaic virus* (ArMV), *Cucumber mosaic virus* (CMV) and *Cherry leaf roll virus* (CLRV) [14–17]. A good alternative for the production of high-quality planting material is in vitro propagation of virus indexed rhubarb plants. Moreover, tissue culture of rhubarb can be a source of bioactive compounds. In rhizomes and leaf petioles of *Rheum rhabarbarum*, the accumulation of catechin, gallic acid, p-cumaric acid, rosmarinic acid, isoquercitrin and resveratrol has been reported [18]. Several articles have been published on the vitro propagation of different culinary rhubarb selections [18,19] and cultivars, including 'Victoria' [20], 'Big Red' [16] and 'Karpov Lipskiego' [21]. For many genotypes, bacteria contaminations, slow culture establishment, low activity of axillary buds and hyperhydricity have been reported [16,18,20]. Mentioned difficulties caused that commercial micropropagation of some valuable rhubarb selections, including 'Malinowy', has not been attained.

The aim of this study was to develop an efficient in vitro propagation technology for the selection of rhubarb 'Malinowy'. Identification and quantification of anthocyanins and soluble sugars by the HPLC method in shoot cultures and ex vitro established plantlets were performed. The development of procedures for rapid in vitro clonal propagation of value rhubarb genotypes may be of great commercial value to the rhubarb industry.

2. Results

2.1. Culture Initiation

Shoot cultures were initiated from axillary buds isolated from dormant, eight-year-old rhizomes. We obtained 51.9–67.6% uncontaminated explants that developed shoots. Contamination of initial explants of *Rheum* 'Malinowy' partly depended on the sanitary status of the mother plant (Table 1, Figure 1A). In MS medium without growth regulators, no explant developed shoots (Table 2). Shoot tips and axillary buds cultured in MS media supplemented with BAP and GA_3 or NAA started to develop into single shoots in 2–3 weeks (Figure 1B). After 3–4 subcultures in the same fresh medium, new shoot formation was observed (Figure 1C). The best growth and development of initial shoots from dormant buds was observed in the modified MS medium containing 75% nitrogen salts, 4.4 μM BAP and 0.3 μM GA_3 (Table 2). After three subcultures in media containing

BAP + NAA, the developed shoots of 'Malinowy' were characterized by pale green leaves and a tendency to hyperhydricity, some of which died. The negative effect of auxin was enhanced when full-strength MS medium was used. It was observed that lowering the strength of nitrogen salts by one quarter in MS medium containing BAP + GA_3 slightly enhanced new shoot formation from initial explants (Table 2).

Table 1. Effect of mother (donor) plants on contamination of initial explants of Polish rhubarb 'Malinowy' in vitro.

Donor Plants	Total Number of Explants	Contaminated Explants (%)		Uncontaminated Explants that Did Not Developed Shoots (%)	Uncontaminated Explants that Developed Shoots (%)
		Bacteria	Fungi		
M1 *	34	29.4	-	3.0	67.6
M2	56	28.6	10.7	8.8	51.9

* 'Malinowy'.

Figure 1. Initiation and stabilization stage of culinary rhubarb 'Malinowy' micropropagation: (**A**) 8-year-old mother plant collected from the plantation at the beginning of November; (**B**) initial shoot developed from axillary bud after 3 weeks of culturing in initiation medium; (**C**) the multiplied shoots at the stabilization stage.

Table 2. Influence of nitrogen salts (75% and 100% MS medium) and growth regulators on the survival rate and initial shoot formation of rhubarb 'Malinowy' after 3 subcultures in the same fresh medium.

Propagation Medium	Survival Rate (%)	Shoot Number/Explant
100% MS control (hormone free)	0	-
100% MS + 4.4 μM BAP + 0.3 μM GA_3	95.3	1.69 bc [1]
100% MS + 4.4 μM BAP + 0.1 μM NAA	58.7	1.07 a
75% MS + 4.4 μM BAP + 0.3 μM GA_3	94.1	1.92 c
75% MS + 4.4 μM BAP + 0.1 μM NAA	72.0	1.31 ab

[1] Means marked with the same letter do not differ significantly ($p = 0.05$) according to Duncan's test; the lowest value is marked with "a".

2.2. *Shoot Multiplication*

The results of our study showed that adenine-type cytokinin (BAP and mT) was effective in the stimulation of shoot formation of rhubarb 'Malinowy' in vitro. It was observed that BAP was the most effective in the stimulation of shoot formation when applied at a concentration of 6.6 μM and *meta*-topolin at 12 μM (Figure 2). The next experiment showed the cytokinin effect on rhubarb shoot formation and quality significantly depended on sucrose supply. The highest multiplication rate (4.8–4.9 shoots/explant) was observed in medium with the lowest sucrose content (59 mM) and in the presence of 6.6 μM BAP or 12.4 μM mT, respectively. Increasing the sucrose supply from 59 to 175 mM in the cytokinin medium decreased the production of shoots and outgrowth of leaves by 3-fold but enhanced shoot length, single shoot mass and callus formation at the base of shoots (Table 3). Progression of shoots from a juvenile to adult phase (shoots with large, dark-green leaf petioles, single root) was also observed (Figure 3A). At 117 mM sucrose in the medium, leaf petioles started to turn red, and increased sucrose supply intensified the color of rhubarb leaf petioles under in vitro conditions. Despite the highest shoot formation rate of rhubarb

'Malinowy' in medium with low sucrose content (59 mM), cyclic multiplication, especially in the presence of mT, resulted in the formation of pale green shoots, with small leaf blades and a tendency to hyperhydricity and deformation. On the other hand, *meta*-topolin stimulated more juvenile shoot formation at higher sucrose levels compared to BAP. This resulted in higher shoot production at 88 mM sucrose and higher shoot quality at high exogenous sucrose supply. Shoots produced in BAP medium with high sucrose content (175 mM) showed a tendency to leaf yellowing and swelling shoot bases (Figure 3B).

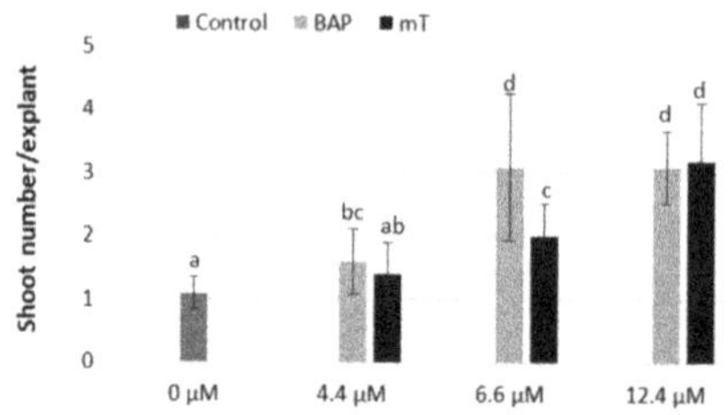

Figure 2. Effect of different cytokinin types (BAP, mT) and concentrations (4.4, 6.6, 12.4 µM) on shoot formation of rhubarb 'Malinowy' after a 4-week subculture period. Sucrose concentration in the medium was 88 mM. Medium without cytokinin was the Control. Means marked with the same letter do not differ significantly ($p = 0.05$) according to Duncan's test.

Table 3. Effect of cytokinin type (BAP, mT) and sucrose levels on shoot formation, fresh mass of shoots, leaf number and shoot length of rhubarb 'Malinowy' after a 4-week subculture period.

Cytokinin Type	Sucrose Concentration (mM)	Fresh Mass (g)			Shoot Number	Shoot Length (mm)	Leaf Number/ Clump
		Shoot Clump	Callus	Single Shoot			
BAP	59	1.90 bc [1]	0.170 a	0.395 a	4.8 e	31.6 a	25.1 d
	73	2.37 cd	0.162 a	0.566 b–c	4.3 de	37.5 bc	21.5 d
	88	1.68 ab	0.261 a–c	0.586 bc	3.1 bc	38.5 bc	16.9 c
	117	1.28 a	0.362 b–d	0.653 c–d	2.2 ab	42.4 cd	11.5 b
	175	1.41 a	0.532 e	0.907 e	1.5 a	49.0 e	8.9 ab
mT	59	2.51 d	0.256 a–c	0.446 ab	4.9 e	30.9 a	32.1 e
	73	2.31 d	0.237 ab	0.522 a–c	4.7 e	36.6 b	24.6 d
	88	1.61 ab	0.423 de	0.494 a–c	3.7 cd	40.1 b–d	12.7 b
	117	1.31 a	0.402 c–e	0.576 b–c	2.3 ab	38.9 bc	11.7 b
	175	1.28 a	0.430 de	0.768 de	1.7 a	44.8 de	6.9 a

[1] Means indicated with the same letter within cytokinin type and sucrose levels do not differ significantly ($p = 0.05$) according to Duncan's test; the lowest value is marked with "a".

2.3. Soluble Sugar and Anthocyanin Contents in Leaf petioles

Among tested soluble sugars, fructose and glucose were dominant in leaf petioles of rhubarb 'Malinowy' after a 4-week subculture period in multiplication medium. Sorbitol was not detected in rhubarb 'Malinowy'. The content of soluble sugars in leaf petioles varied significantly depending on sucrose concentration in the medium and cytokinin type. Increasing sucrose concentration (from 59 to 175 mM) resulted in an increase in endogenous soluble sugar content by 60% and 75% in the presence of BAP and mT, respectively. Generally, BAP influenced higher fructose and glucose accumulation compared to *meta*-topolin, especially at the lowest sucrose level in the medium. In medium with BAP and high sucrose content (175 mM), the accumulation of a small amount of sucrose was also observed (Figure 4).

Figure 3. In vitro shoot formation, rooting and acclimatization of culinary rhubarb 'Malinowy'. (**A**) Shoots multiplicated in mT medium containing different levels of sucrose (from the left: 59, 73, 88 and 175 mM); (**B**) shoots from BAP medium containing 175 mM sucrose; (**C**) leaf petioles from mT medium containing 175 mM sucrose; (**D**) shoot cultures; (**E**) roots three weeks after transfer to medium with 0.49 μM IBA; (**F**) roots three weeks after transfer to medium with 4.9 μM IBA; (**G**) shoots rooted in the presence of 4.9 μM IBA; (**H**) roots after 4 weeks in medium with 4.9 μM IBA; (**I–J**) plantlets four weeks after transfer to ex vitro conditions (growth room); (**K–L**) plantlets six weeks after transfer to the greenhouse; (**M–O**) plantlets nine weeks after transfer to the greenhouse; (**N–P**) leaf petioles after different periods of growing ex vitro.

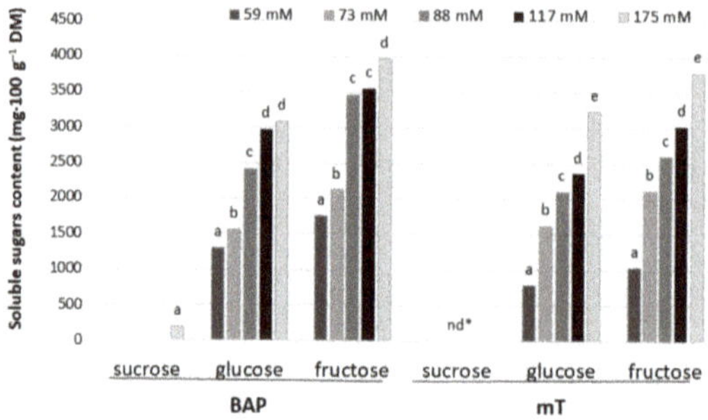

Figure 4. Contents of soluble sugars in culinary rhubarb 'Malinowy' after a 4-week subculture period in medium containing different cytokinin types: BAP (6.6 µM) and mT (12.4 µM) and sucrose levels (59, 73, 88, 117 and 175 mM); * not detectable. Duncan's test was used independently for each type of cytokinin and sucrose concentration. Different letters indicate significant differences among sucrose treatments ($p = 0.05$).

We found that exogenous sucrose levels and cytokinin type had a significant effect on anthocyanin accumulation in rhubarb 'Malinowy' under in vitro conditions (Figures 3A and 5). The most abundant anthocyanin compound was cyanidin-3-O-rutinoside (max. 208.2 mg·100 g^{-1} DM), followed by cyanidin-3-O-glucoside (max. 47.7 mg·100 g^{-1} DM). Both cyanidin levels were enhanced in rhubarb leaf petioles by increased sucrose concentration in the medium. At a high sucrose level (175 mM), the production of anthocyanins was significantly stimulated by cytokinin. It was found that *meta*-topolin resulted in 50% higher content of anthocyanins in rhubarb petioles compared to BAP (Figure 5).

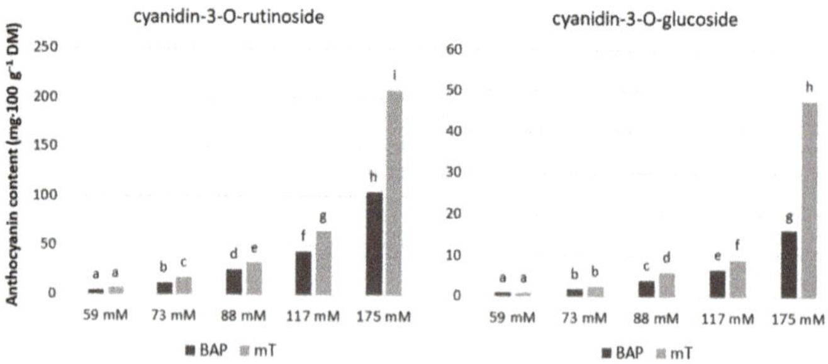

Figure 5. Contents of anthocyanin in culinary rhubarb 'Malinowy' after a 4-week subculture period in medium containing different cytokinin types: BAP (6.6 µM) and mT (12.4 µM) and sucrose levels (59, 73, 88, 117 and 175 mM). Means indicated with the same letter within cytokinin treatment do not differ significantly ($p = 0.05$) according to Duncan's test.

2.4. Rooting and Acclimatization

Under in vitro conditions, roots emerged 1–2 weeks after transfer to the rooting media. As shown in Table 4, rooting of culinary rhubarb 'Malinowy' was significantly affected by IBA levels. In auxin-free medium, roots formed with the effectiveness of 40%. Application of IBA resulted in an increase in root formation in a concentration-dependent manner (Table 4). The best rooting response (94.9% rooting frequency; 10.7 roots/shoot) was observed in the presence of IBA at the concentration of 4.9 µM (Figure 3F). In all treatments, root length progressively increased with time (Figure 3F) and resulted in their damage during transfer to ex vitro conditions (Figure 3H).

Table 4. Rooting response of rhubarb 'Malinowy' shoots after three-week growth in MS medium without auxin (control) and supplemented with IBA at different concentrations (0.49, 2.5, 4.9 µM).

Treatment	Auxin Concentrations (µM)	Rooting Frequency (%)	Root Number/Shoot	Root Length (mm)	Shoot Length (mm)
Control	0.0	40.0 a [1]	2.1 a	29.2 b	75.2 a
	0.49	55.9 b	2.5 a	28.8 b	72.0 a
IBA	2.5	82.0 c	2.6 a	30.3 b	77.8 a
	4.9	94.9 d	10.7 b	13.84 a	78.5 a

[1] Means of each parameter marked with the same letter do not differ significantly (p = 0.05) according to Duncan's test; the lowest value is marked with "a".

We then compared the efficiency of acclimatization of rhubarb shoots rooted in vitro and directly ex vitro. The development of the root system was evaluated on a three-point scale (Figure 6). It was observed that in vitro rooted shoots rapidly developed a root system (Table 5, Figure 3J). Three weeks after transfer to ex vitro conditions, roots were visible through the walls of the peat plugs. This resulted in more vigorous growth of shoots and higher ability to tolerate low humidity compared to shoots directly rooted and acclimatized. After four weeks of acclimatization, in vitro rooted shoots were transferred to the greenhouse. Ex vitro rooted shoots needed at least 7–8 weeks to develop a root system and acclimatize. However, after this period of time, the rooting rate was poorer than after 4-week acclimatization of shoots rooted in vitro. The ex vitro establishment rate of in vitro rooted shoots was 100%, while those rooted ex vitro survived with 83%.

Figure 6. Development of the root system of culinary rhubarb 'Malinowy' on a 1–3 scale.

Table 5. Effect of microcutting status (rooted, unrooted) on survival, shoot growth and root system development of *Rheum* 'Malinowy' after 4 weeks of growth ex vitro (growing room).

Type of Microcutting	Survival Frequency (%)	Cutting Length (mm)	Rooting Index
Unrooted	83	72.8 a [1]	1.9 a
Rooted	100	107.0 b	2.9 b

[1] Means of each parameters indicated with the same letter do not differ significantly (p = 0.05) according to Duncan's test; the lowest value is marked with "a".

After transfer to the greenhouse, growth of the plantlets tended to decrease, and yellowing of the oldest leaves was observed. As shown, soluble sugar and anthocyanin contents in leaf petioles were very low at this stage of growth in the greenhouse (Table 6). On sunny days, the plantlets needed shading. After two weeks, new leaves started to develop. Six weeks after transfer to the greenhouse, the average length of leaf petioles was enhanced by 15.6% and leaf area by 47.8%. A rapid increase in leaf (petioles and blades) thickness was also observed (Figure 3K). Leaf petioles of all rhubarb plantlets turned red (Figure 3M–N). During six-week growth in the greenhouse, contents of anthocyanins and soluble sugars in the leaf petioles were enhanced by 4 and 7 times, respectively. They contained sucrose, glucose and fructose in a ratio 1:2:2. Total anthocyanin content in leaf petioles of 'Malinowy' was 188 mg·100 g^{-1}, including 94% cyanidin-3-O-rutinoside and 6% cyanidin-3-O-glucoside (Table 6). No morphologically aberrant plantlets were found.

The ten-week-old rhubarb plantlets were transferred (at the end of April) to the plantation for further observation.

Table 6. Morphological and biochemical characteristics of *Rheum* 'Malinowy' during a 6-week growth period in the greenhouse.

Growth Duration	Length of Leaf Petioles (mm)	Leaf Area (cm^2)	Soluble Sugar Contents ($mg \cdot 100\ g^{-1}$ DM)	Anthocyanin Contents ($mg \cdot 100\ g^{-1}$ DM)
1 week	60.5	20.2	Sucrose-t.a. Glucose-762 Fructose-858	cyanidin-3-O-glucoside-2.58 cyanidin-3-O-rutinoside-45.5
6 weeks	71.1	38.8	Sucrose-2196 Glucose-4958 Fructose-4790	cyanidin-3-O-glucoside-11.3 cyanidin-3-O-rutinoside-176.7

t.a.—trace amounts.

3. Discussion

The quality and quantity of rhubarb crops are significantly dependent on planting material. Tissue cultures are a useful tool for vegetative propagation of virus-free rhubarb plants and rapid multiplication of valuable selections with the highest possible content of bioactive ingredients. For commercial micropropagation of planting material, a successful in vitro propagation protocol including all stages (initiation of aseptic culture, shoot multiplication, rooting of microshoots, ex vitro acclimatization) is very important.

Axillary bud development has proven to be the most often applied system for producing true-to-type plantlets. In the present study, dormant axillary buds of culinary rhubarb 'Malinowy' were used successfully for the initiation of shoot cultures. By using a two-step disinfection procedure, we obtained an average of 59.5% uncontaminated explants from eight-year-old rhizomes that developed shoots of high quality. Given the age of the rhizomes and the associated large amount of dead tissue and secreted mucus, as well as the presence of endogenous bacteria, this is a high efficiency of culture establishment. Buds of 'Big Red' treated with 2% sodium hypochlorite for 15 min showed 51% establishment success. Additional methods for disinfection in sodium dichloroisocyanurate (300 $mg \cdot L^{-1}$ for 20 min and 48 h) and 4% chlorine dioxide did not reduce the amount of contamination [16]. Clapa et al. [18], by using bleach solution of 20% (ACE-Protector) for 20 min, obtained 20% contaminated explants, but 36% were necrotic and only 44% were viable.

Data in the literature indicate that initial growth of culinary rhubarb shoots from axillary buds depends on different factors, including growth regulators, term of explant collecting, bud size and genotype [16–21]. The most important factor stimulating the activity of rhubarb buds was BAP added singly [16] or in combination with auxin [19,20]. We obtained the best growth response of dormant buds of rhubarb 'Malinowy' in modified MS medium containing 75% nitrogen salts, 4.4 μM BAP and 0.3 μM GA_3. The combination of cytokinin and GA_3 was found to increase the activity of axillary buds of many plant species, especially perennials and woody plants [22–24]. On the other hand, we found that NAA presence in the initial medium significantly decreased shoot quality. This might be related to the inappropriate type and concentration of auxin or lack of need of auxin for rhubarb 'Malinowy' shoot induction. We also observed that the negative effect of auxin was enhanced when full-strength MS medium was used. Murashige and Skoog medium with KNO_3 (at 1900 $mg \cdot L^{-1}$) and NH_4NO_3 (at 1650 $mg \cdot L^{-1}$) is the most common medium for micropropagation of many horticultural plants, but these high concentrations of nitrogen salts are supraoptimal for some plant species and can induce different physiological disorders, including hyperhydricity [25]. Ogura-Tsujita and Okuba [26] reported that rhizome explants of *Cymbidium kanran* growing in low-nitrogen medium produced 55% less ethylene than those coincident with enhanced shoot production.

It has been reported that among the cytokinins tested, BAP was more effective than isopentenyladenine (2iP), kinetin and thidiazuron (TDZ) for axillary multiplication of various culinary rhubarb genotypes [19,21]. The optimal BAP concentration ranged from

2.2 to 22.2 μM according to rhubarb cultivar [16,20]. Kozak and Sałata [21] showed that 'Karpow Lipskiego' cultured in BAP medium was characterized by the largest number of leaves, basal tissue and fresh mass of shoots. However, the use of kinetin and 2iP enhanced the length of shoots. The genotype-dependent multiplication rate and shoot morphology of some culinary rhubarb cultivars and clones have already been observed [16]. For example, rhubarb 'Malinowy' cultured in BAP medium formed much lower shoots but with an enhanced number of leaves per explant compared to 'Karpow Lipskiego' [21]. Similar to a study on *Rheum rhabarbarum* [18], we observed that *meta*-topolin was of similar efficiency in the stimulation of axillary multiplication of rhubarb 'Malinowy' compared to BAP. Additionally, our study showed that both cytokinins stimulated shoot formation of rhubarb in a manner dependent on sucrose concentration. Increasing sucrose supply (from 59 to 175 mM) in the growing medium resulted in a reduction in the multiplication rate by 70% and the stimulation of mature shoot production. Sucrose inhibition of shoot formation was previously observed in different plant species, including *Helleborus niger* L., *Paeonia lactiflora* Pall, [24,27], *Pelargonium hortorum* L.H. Bailey [28] and *Rosa* 'Konstancin' [29].

It is known that sugars play important roles in in vitro cultures as energy and carbon sources and osmotic agents. They can also act as signaling molecules and/or as regulators of gene expression [30]. Sugar-mediated signals indicate carbohydrate availability and regulate metabolism with sugar usage and storage [31]. As shown in our study, the exogenous sucrose supply in cytokinin medium affected the accumulation of fructose and glucose in rhubarb leaf petioles and coincided with a reduced multiplication rate and plantlet transition from juvenile to mature phase. At high exogenous sucrose content in the medium, BAP influenced the formation of more mature shoots compared to mT. Moreover, our study showed that high exogenous sucrose content promotes the accumulation of anthocyanins in leaf petioles of rhubarb in vitro.

Anthocyanins are an important class of flavonoids that represent a large group of plant secondary metabolites. They are recognized for their diverse function in plant development and beneficial effect on human health. Generally, anthocyanin accumulation in fruits and vegetables are accompanied by their maturation and is regulated by both developmental and environmental cues [32]. Sugar-induced anthocyanin accumulation has been observed in many plant species. In *Petunia* Juss, sugars were shown to be required for the pigmentation of developing corollas [33], while in grape berry skin, sugars were found to induce most genes involved in anthocyanin synthesis [34]. Enhanced anthocyanin accumulation caused by high sucrose content in the medium has been previously reported in *Melastoma malabathricum* L. [35], *Clematis pitcheri* [36] and *Petunia* [37]. It is well known that phytohormones can also modulate anthocyanin biosynthesis by regulating the expression of genes involved in the flavonoid biosynthetic pathway [38,39]. As shown in our study, the sucrose-induced accumulation of anthocyanin in rhubarb petioles in vitro was stimulated by cytokinin, mainly *meta*-topolin. Potential of mT in stimulating the accumulation of proanthocyanidins has been previously observed for the shoot cultures of *Musa* 'William' [40]. The authors demonstrated that *meta*-topolin was the most effective in this process compared to other topolins and BAP.

During the last decade, *meta*-topolin has been widely used in micropropagation of many plant species, improving multiplication and rooting efficiency [41–43]. Our previous observation showed that the shoot of rhubarb 'Malinowy' treated with *meta*-topolin had higher rooting capacity than those cultured in BAP medium before rooting (data not shown). It was shown that the rooting ability of rhubarb in vitro is also dependent on genotype. For example, 'Karpow Lipskiego' produced 100% rooted shoots in hormone-free medium and spontaneously formed multiple roots in multiplication medium [21]. We obtained only 40% rooted shoots in auxin-free medium, but the application of IBA at 4.9 μM resulted in 100% rooting frequency and multiple root formation. Thomas [16] reported 86% rooting frequency for 'Big Red' in the presence of 2.9 μM 3-indoleacetic acid (IAA). Our study results are in agreement with those by Thomas [16], showing that rhubarb shoots rooted in vitro showed better shoot growth and higher survival rate after transfer to ex vitro conditions compared to direct rooting and acclimatization.

We found plantlets of 'Malinowy' showed a rapid increase in leaf size, petiole length and diameter when they were transferred to the greenhouse. Anthocyanin analyses revealed that ten-week old 'Malinowy' plantlets contained a very high amount of cyanidyn-3-O-rutinoside (176.7 mg·100 g^{-1}) and cyanidin-3-O-glucoside (11.3 mg·100 g^{-1}). Cyanidin derivatives were found previously to be the main anthocyanin in rhubarb stalk grown in the field. It has been demonstrated that rhubarb cultivars significantly differ in their anthocyanin content and percentage of the two main cyanidins [7,9]. The study presented by Takeoka et al. [9] showed that of twenty-one cultivars, total anthocyanin content ranged from 19.8 ('Crimson Red') to 341.1 mg·100 g^{-1} ('Valentine'). Similarly, as in our study, cyanidyn-3-O-rutinoside was the main anthocyanin present in culinary rhubarb 'Red Malinowy' [11]. Moreover, in this genotype, twenty other phenolic compounds, including flavan-3-ols, flavonols and gallotannins, were identified. Among them, the most abundant was catechin (112 ng·mg^{-1} dry mass). The study of Clapa et al. [18] showed that phenolic composition can differ between in vitro and field-grown *Rheum rhabarbarum* plant extract. Further observations and analyses of phytochemical and nutritional compounds for rhubarb 'Malinowy' plants growing in vitro and in the field are needed.

4. Materials and Methods

4.1. Plant Material

Two donor plants of Polish garden rhubarb 'Malinowy' were carefully selected from a plantation in Kańczuga, located in the Subcarpathia Province, in southeastern Poland (WGS-84: 49°59′0″ N, 22°24′42″ E). Among other plants, they were distinguished by high yield and straight, thick and intensive red stalks. The selected plants were harvested at the beginning of November. Before culture initiation, the donor plants were tested by enzyme-linked immunosorbent assay (DAS-ELISA) with commercially available antibodies against ArMV, TuMV, *Cucumber mosaic virus* (CMV), *Cherry leaf roll virus* (CLRV), *Tobacco ring spot virus* (TRSV), *Tomato ring spot virus* (ToRSV), *Tomato spotted wilt virus* (TSWV), *Tomato black ring virus* (TBRV), *Strawberry latent ringspot virus* (SLRV) (LOEWE Biochemica, Sauerlach, Germany) and *Tobacco mosaic virus* (TMV) (Agdia, Elkhart, IN) by the method of Clark and Adams [44]. The plants were found to be virus free.

4.2. Culture Initiation

Shoot cultures were established from axillary buds isolated from dormant, eight-year-old rhizomes. First, they were divided into small parts then washed thoroughly with running tap water and soaked in fungicide. Buds were isolated and sterilized by soaking in commercial bleach (ace 4 mL/water 100 mL) for 20 min and then in 0.1% $HgCl_2$ for 5 min. After rinsing in sterile water, the explants were placed in 50 mL Erlenmeyer flasks in Murashige and Skoog [45] (MS) medium containing different levels of nitrogen salts (100% and 75%), 100 mg·L^{-1} myo-inositol, vitamins (nicotinic acid, pyridoxine, thiamine (1.0 mg·L^{-1} each)), 2 mg·L^{-1} glycine, 88 mM sucrose and 6 g·L^{-1} agar (Biocorp, Poland). The effect of 4.4 µM benzylaminopurine (BAP) added together with 0.3 µM gibberellic acid (GA_3) or 0.1 µM 1-naphthaleacetic acid (NAA) was studied. Full-strength MS medium without growth regulators was the control. The pH of the medium was adjusted to 5.8 before autoclaving. After 3 subcultures (each 3 weeks), the survival rate and the number of developed shoots were determined.

The shoots in all in vitro experiments were maintained at the temperature of 20 ± 2 °C under a standard 16/8 h photoperiod of 40 µmol m^{-2} s^{-1} (warm-white fluorescent lamps).

4.3. Shoot Multiplication

Murashige and Skoog medium with nitrogen salts reduced by a quarter containing 100 mg·L^{-1} myo-inositol, vitamins (nicotinic acid, pyridoxine and thiamine (1.0 mg·L^{-1} each) and 2 mg·L^{-1} glycine was used throughout the experiments. To obtain effective, cyclic shoot multiplication of rhubarb 'Malinowy', first, the effect of two adenine-type cytokinins (BAP and hydroxybenzylaminopurine (*meta*-topolin (mT)), added in concentrations of 4.4 μM, 6.6 μM and 12.4 μM, was examined (Experiment 1). After two subcultures (each lasting 4 weeks), the multiplication rate was determined. Then, the interaction between cytokinins (BAP or mT) and sucrose added at different concentrations (59, 73, 88, 117 and 175 mM) was studied. In Experiment 2, BAP was used at 6.6 μM and mT at 12.4 μM. After two subcultures, fresh mass, number and length of shoots, and number of leaves were determined. Leaf petioles were collected, lyophilized, and crushed into a homogenous powder using a laboratory mill (A 11, IKA, Staufen, Germany). The samples were then subjected to qualitative and quantitative analyses of sugars and anthocyanins by the HPLC method.

4.4. Rooting and Acclimatization

The aim of the experiments was to compare the efficacy of in vitro and ex vitro rooting and acclimatization. In the last subculture before rooting, shoots were grown in medium containing 4.1 μM *meta*-topolin.

For in vitro rooting, the selected shoots approximately 4 cm long were placed in modified MS medium containing 88 mM of sucrose and indole-3-butyric acid (IBA) at different concentrations (0, 0.49, 2.5 and 4.9 μM). After 3 weeks, rooting frequency, number of roots per explant and length of roots were determined. For the acclimatization experiment, shoots containing at least 5–6 roots were selected.

For ex vitro rooting and acclimatization, the shoots were taken directly from mT medium. The bases of shoots (approx. 4 cm long) were dipped in commercial rooting powder (Rhizopon AA 0.5%, Poland). Both types of microcutting (in vitro rooted and unrooted) were planted in multicellular trays 30 mm in diameter with a mixture of peat (Alonet Substrat, SIA Florabalt, LV-5106 Valle, Latvia) and perlite (2:1) in plastic plug boxes covered with transparent plastic caps to prevent dehydration. Ex vitro rooting and acclimatization took place in a growth room (25 ± 2 °C; PPFD—50 μmol m^{-2} s^{-1}). The microcuttings were hardened by gradually decreasing air humidity. After 10 days, they were fed with 0.1% Kristalon (Yara Vlaardingen B.V., Netherlands) containing 18:18:18 (*w/w/w*) NPK. Four weeks after transfer to ex vitro conditions, the following data were collected: length of plantlets and the visual estimation of the root system on a three-point scale (1—no roots or single, short roots; 2—poorly developed root system, 3—well-developed root system, roots out of trays) (Figure 6).

After four weeks of acclimatization, in vitro rooted shoots were transferred to a greenhouse (at the end of February). They were grown at a temperature of 20/18 °C (day/night) with a 16 h photoperiod provided by additional lighting. Plants were manually watered and fed with 0.1% Kristalon. Plantlet growth (length of leaf petioles and leaf area) and biochemical status (soluble sugars and anthocyanin contents) were assessed after 1 and 6 weeks of growth in the greenhouse.

4.5. Soluble Sugar Analysis

The sugars were quantified by calibration curve for sucrose, glucose, fructose and sorbitol standards in the concentration range of 20–250 mg/100 mL, and the results are expressed as mg·100 g^{-1} dry mass (DM). An example chromatogram of sugar separation is shown in the Supplementary Materials (Figure S1). HPLC analysis of sugars was carried out with the HP 1200 system (Agilent Technologies, Waldbronn, Germany) equipped with an RI Detector with a BioRad Aminex-87C column (300 × 7.5 mm) according to European Standard EN 12630. The mobile phase was water purified by the MiliQ System (Milipore, Molsheim, France), isocratic flow was 0.6 mL·min^{-1} and column temperature was 80 °C. The lyophilized powder (100 mg) was extracted in 4 mL of redistilled water for 20 min in an ultrasonic bath. The suspension was then centrifuged (7000× *g*, 10 min). The resulting extract for sugar determination was filtered through a Sep-Pak® PLUS C18 filter (Waters, Ireland).

4.6. Estimation of Anthocyanins by HPLC

HPLC analysis of anthocyanins was performed according to the method described by Nielsen et al. [46] with some modifications. In short, 5 μL of the eluate was analyzed using an Agilent HPLC Model HP 1200 equipped with a diode array detector (DAD). Separation was performed on a Phenomenex®Fusion RP column (250 mm × 4.6 mm; particle size = 4 μm) using a mobile phase consisting of water/ formic acid (95:5 *v/v*) (A) and acetonitrile (B). Elution profile: 0–16 min, 3%–9% B; 16–30 min, 12% B; 30–54 min, 33% B; 54–58 min, 90% B; 58–62 min, isocratic 90% B. The anthocyanins in the eluate were detected at 520 nm and a temperature of 25 °C. Their amounts were quantified by calibration with the standards of cyanidin-3-O-glucoside and cyanidin-3-O-rutinoside and expressed in mg·100 g^{-1} dry mass (DM). An example chromatogram of anthocyanins separation is shown in Supplementary Material (Figure S2).

The lyophilized powder (50 mg) was extracted in 2 mL of 60% methanol acidified with 1% formic acid for 20 min in an ultrasonic bath. The suspension was then centrifuged (7000× *g*, 10 min). The resulting extract for phenolic compound determination was filtered through a PTFE filter (0.45 μm, 15 mm).

4.7. Statistical Analysis

For the multiplication and in vitro rooting experiments, 30 shoots (6 shoots × 5 glass jars) were used in each treatment. For ex vitro rooting and acclimatization, 50 rooted and 50 unrooted shoots were used. The experiments were carried out twice. The final data were the means of the two replicated experiments. The data were subjected to one- (rooting and acclimatization experiment) or two-factor analysis of variance (ANOVA). The significance of the differences between means was evaluated by Duncan's test at $p = 0.05$.

5. Conclusions

We developed a practical protocol for mass propagation of the selected genotype of rhubarb with a high content of anthocyanins. It enhanced the availability of phenolic-rich planting material for the establishment of commercial rhubarb plantations. The study indicates that growth and development of rhubarb shoots in vitro and secondary metabolite production are modulated by cytokinin and sucrose concentration. It was found that *meta*-topolin is a very useful cytokinin for rhubarb 'Malinowy' in vitro, which can be a good alternative for cultivars revealing multiplication and rooting difficulty in the presence of BAP. Although direct ex vitro rooting is possible, rhubarb shoots rooted in vitro showed better ex vitro establishment and growth in the greenhouse. Hence, in vitro rooting of rhubarb 'Malinowy' is preferred for mass production. Our results indicated that anthocyanins in leaf petioles of rhubarb plantlets were influenced by plant developmental stage and in vitro conditions. It is possible to elicit anthocyanins by high sucrose concentration combined with *meta*-topolin. Finally, the results obtained give important suggestions for introduction into the market and cultivation for rhubarb 'Malinowy', following the latest research conducted globally also on other interesting wild and cultivated plants [47–49].

Supplementary Materials: The following are available online at https://www.mdpi.com/article/10.3390/plants10091768/s1, Figure S1: HPLC chromatograms of: (**A**) sugars in standard solution; (**B**) sugars in rhubarb 'Malinowy'. Peak identification: 1—sucrose (Rt −8.1 min.), 2—glucose (Rt −9.8 min), 3—fructose (Rt −12.7 min), 4—sorbitol (Rt −21.4 min.). Figure S2: HPLC chromatograms of: (**A**) antocyanins in standard solution; (**B**) antocyanin in rhubarb 'Malinowy'. Peak identification: 1—cyanidin-3-O-glucoside (Rt −22.3 min.), 2—cyanidin-3-O-rutinoside (Rt −25.7 min); Rt—retention time.

Author Contributions: A.W.—designed and performed all experiments, collected and analyzed data, prepared figures and tables, writing—review and editing of the paper M.M.-F.—performed analyses of soluble sugars and anthocyanins, collected data and revised the manuscript. All authors have read and agreed to the published version of the manuscript.

Funding: This research was funded by the European Union under the "European Agricultural Fund for Rural Development: Europe investing in rural areas", Grant Number 00009.DDD.6509.00020.2019.09.

Institutional Review Board Statement: Not applicable.

Informed Consent Statement: Not applicable.

Data Availability Statement: The data presented in this study are available on request from the corresponding author.

Acknowledgments: The authors thank Prof. Mirosława Cieślińska for virus analyses and Prof. Małgorzata Podwyszyńska for kindly revising this manuscript.

Conflicts of Interest: The authors declare no conflict of interest.

References

1. Wang, A.; Yang, M.; Liu, J. Molecular phylogeny, recent radiation and evolution of gross morphology of the rhubarb genus *Rheum* (*Polygonaceae*) inferred from chloroplast DNA *trn*L-F sequences. *Ann. Bot.* **2005**, *96*, 489–498. [CrossRef] [PubMed]
2. Zheng, Q.; Wu, H.F.; Guo, J.; Nan, H.J.; Chen, S.L.; Yang, J.S.; Xu, X.D. Review of rhubarbs: Chemistry and pharmacology. *Chin. Herb. Med.* **2013**, *5*, 9–32.
3. Xiang, H.; Zuo, J.; Guo, F.; Dong, D. What we already know about rhubarb: A comprehensive review. *Chin. Med.* **2020**, *15*, 88. [CrossRef] [PubMed]
4. Bratsch, A.; Mainville, D. *Specialty crop profile: Rhubarb*; Virginia Tech: Blacksburg, VA, USA, 2009; Publication 438–110.
5. Foust, C.M.; Marshall, D.E. Culinary rhubarb production in North America: History and recent statistics. *HortScience* **1991**, *26*, 1360–1363. [CrossRef]
6. Roberfroid, M.B. Concepts and strategy of functional food science: The European perspective. *Am. J. Clin. Nutr.* **2000**, *71*, 1660S–1664S. [CrossRef] [PubMed]
7. Oszmiański, J.; Wojdyło, A. Polyphenol content and antioxidative activity in apple purées with rhubarb juice supplement. *Int. J. Food Sci. Technol.* **2008**, *43*, 501–508. [CrossRef]
8. Krafczyk, N.; Kötke, M.; Lehnert, N.; Glomb, M.A. Phenolic composition of rhubarb. *Eur. Food Reseach Technol.* **2008**, *228*, 187–196. [CrossRef]
9. Takeoka, G.R.; Dao, L.; Harden, L.; Pantoja, A.; Kuhl, J.C. Antioxidant activity, phenolic and anthocyanin contents of various rhubarb (*Rheum* spp.) varieties. *Int. J. Food Sci. Technol.* **2013**, *48*, 172–178. [CrossRef]
10. Kołodziejczyk-Czepas, J.; Liudvytska, O. *Rheum rhaponticum* and *Rheum rhabarbarum*: A review of phytochemistry, biological activities and therapeutic potential. *Phytochem. Rev.* **2021**, *20*, 589–607. [CrossRef]
11. Kalisz, S.; Oszmiański, J.; Kolniak-Ostek, J.; Grobelna, A.; Kieliszek, M.; Cendrowski, A. Effect of a variety of polyphenols compounds and antioxidant properties of rhubarb (*Rheum rhabarbarum*). *Food Sci. Technol.* **2020**, *118*, 108775. [CrossRef]
12. Kuhl, J.C.; DeBoer, V.L. Genetic diversity of rhubarb cultivars. *J. Am. Soc. Hortic. Sci.* **2008**, *133*, 587–592. [CrossRef]
13. Maynard, D.N. Evaluation of propagating stock and practices for annual rhubarb production. *HortScience* **1990**, *25*, 1399–1401. [CrossRef]
14. Tomlinson, J.A.; Walkey, D.G.A. The isolation and identification of rhubarb viruses occurring in Britain. *Ann. Appl. Biol.* **1976**, *59*, 415–427. [CrossRef]
15. Robertson, N.L.; Ianson, D.C. First report of *Turnip mosaic virus* in rhubarb in Alaska. *Plant Dis.* **2005**, *89*, 430. [CrossRef]
16. Thomas, J. *Virus Identification and Development of Long-Term Management Strategies for the Rhubarb Industry*; Final report Hal Project VG05053; Horticulture Australia: Sydney, NSW, Australia, 2011; p. 93.
17. Komorowska, B.; Ptaszek, M.; Jarecka-Boncela, A.; Hasiów-Jaroszewska, B. First report of Arabis Mosaic Virus in Rhubarb in Poland. *Plant Dis.* **2018**, *102*, 1863. [CrossRef] [PubMed]
18. Clapa, D.; Borsai, O.; Hârta, M.; Bonta, V.; Szabo, K.; Coman, V.; Bobis, O. Micropropagation, genetic fidelity and phenolic compound of *Rheum rhabarbarum* L. *Plants* **2020**, *9*, 656. [CrossRef] [PubMed]
19. Lepse, L. Adapting technology for local rhubarb (*Rheum rhaponticum* L.) clones propagation In Vitro. *Latv. J. Agron.* **2005**, *8*, 324–327.
20. Lassus, C.; Voipio, I. Micropropagation of rhubarb with special reference to weaning stage and subsequent growth. *Agric. Sci. Finl.* **1994**, *3*, 189–194. [CrossRef]
21. Kozak, D.; Sałata, A. Effect of cytokinins on In Vitro multiplication of rhubarb (*Rheum rhaponticum* L.) 'Karpov Lipskiego' shoots and ex vitro acclimatization and growth. *Acta Sci. Pol. Hortorum Cultus* **2011**, *10*, 75–87.
22. Beretta, D.; Eccherr, T. Determination of optimal level of BAP, GA_3 and NAA in micropropagation of *Camellia* by an orthogonal composite design. *Acta Hortic.* **1987**, *212*, 151–154. [CrossRef]
23. Wojtania, A.; Skrzypek, E.; Gabryszewska, E. Morphological and biochemical responses to gibberellic acid in *Magnolia* × 'Spectrum' In Vitro. *Acta Biol. Crac. Ser. Bot.* **2016**, *58*, 103–111. [CrossRef]
24. Gabryszewska, E.A. Effect of different sucrose and nitrogen salt levels in the medium and temperature on In Vitro propagation of *Helleborus niger* L. *Acta Agrobotanica* **2015**, *68*, 161–171. [CrossRef]

25. El-Dawayati, M.M.; Zayed, Z.E. Controlling hyperhydricity in date palm In Vitro culture by reduced concentration of nitrate nutrients. In *Date Palm Biotechnology Protocols Volume I: Tissue Culture Applications*; Al-Khayri, J., Jain, S., Johnson, D., Eds.; Humana Press: New York, NY, USA, 2017; Volume 1637. [CrossRef]
26. Ogura-Tsujita, Y.; Okuba, H. Effects of low nitrogen medium on endogenous changes in ethylene, auxins, and cytokinins in In Vitro shoot formation from rhizomes of Cymbidium kanran. *Vitr. Cell. Dev. Biol.-Plant* **2006**, *42*, 614–616. [CrossRef]
27. Gabryszewska, E. The effect of glucose and growth regulators on the organogenesis of *Paeonia lactiflora* Pall. in vitro. *J. Fruit Ornam. Plants Res.* **2010**, *18*, 309–320.
28. Wojtania, A.; Węgrzynowicz-Lesiak, E.; Dziurka, M.; Waligórski, P. Sucrose and cytokinin interactions in relations to ethylene and abscisic acid production in the regulation of morphogenesis in *Pelargonium* × *hortorum* L.H. BAILEY In Vitro. *Acta Biol. Crac. Ser. Bot.* **2015**, *57*, 62–69. [CrossRef]
29. Wojtania, A.; Matysiak, B. In vitro propagation of Rosa 'Konstancin' (*R. rugosa* × *R. beggeriana*), a plant with high nutritional and pro-health value. *Folia Hortic.* **2018**, *30*, 259–267. [CrossRef]
30. Smeekens, S.; Ma, J.; Hanson, J.; Rolland, F. Sugar signal and molecular networks controlling plant growth. *Curr. Opin. Plant Biol.* **2010**, *13*, 274–279. [CrossRef]
31. Ciereszko, I. Regulatory roles of sugars in plant growth and development. *Acta Soc. Bot. Pol.* **2018**, *87*, 3583. [CrossRef]
32. Liu, Y.; Tikunov, Y.; Schouten, R.E.; Marcelis, F.M.; Visser, R.G.F.; Bovy, A. Antocyanin biosynthesis and degradation mechanisms in Solanaceous vegetables: A review. *Front. Chem.* **2018**, *6*, 52. [CrossRef] [PubMed]
33. Weiss, D. Regulation of flower pigmentation and growth: Multiple signalling pathways control anthocyanin synthesis in expanding petals. *Physiol. Plant.* **2000**, *110*, 152–157. [CrossRef]
34. Gallop, R.; Farhi, S.; Peri, A. Regulation of the leucoanthocyanindin dioxygenase gene expression in *Vitis vinifera*. *Plant Sci.* **2001**, *161*, 579–588. [CrossRef]
35. See, K.S.; Bhatt, A.; Keng, C.L. Effect of sucrose and methyl jasmonate on biomass and anthocyanin production in cell suspension culture of *Melastoma malabathricum* (*Melastomaceae*). *Rev. Biol. Trop.* **2011**, *59*, 597–606. [PubMed]
36. Kawa-Miszczak, L.; Węgrzynowicz-Lesiak, E.; Gabryszewska, E.; Saniewski, M. Effect of different sucrose and nitrogen levels in the medium on chlorophyll and anthocyanin content i clematis pitcher shoots cultured In Vitro at different temperatures. *J. Fruit Ornam. Plant Res.* **2009**, *17*, 113–121.
37. Ai, T.N.; Naing, A.H.; Lim, S.-H.; Kim, C.K. Sucrose-induced anthocyanin accumulation in vegetative tissue of Petunia plants requires anthocyanin regulatory transcription factor. *Plant Sci.* **2016**, *252*, 144–150. [CrossRef] [PubMed]
38. Das, P.K.; Shin, D.H.; Choi, S.B.; Park, Y.I. Sugar-hormone cross-talk in anthocyanin biosynthesis. *Malecules Cells* **2012**, *34*, 501–507. [CrossRef] [PubMed]
39. Góraj-Koniarska, J.; Saniewski, M. The effect of sugars in relations to methyl jasmonate on anthocyanin formation in the roots of *Kolanchoe blossfeldiana* (Poelln.). *Acta Agrobot.* **2014**, *68*, 173–178.
40. Aremu, A.O.; Bairu, M.W.; Szüčová, L.; Doležal, K.; Finnie, J.F.; Van Staden, J. Assessment of the role of *meta*-topolins on In Vitro produced phenolics and acclimatization competence of micropropagated 'Williams' banana. *Acta Physiol. Plant.* **2012**, *34*, 2265–2273. [CrossRef]
41. Wojtania, A. Effect of *meta*-topolin on In Vitro propagation of *Pelargonium* × *hortorum* and *Pelargonium* × *hederaefolium* cultivars. *Acta Soc. Bot. Pol.* **2010**, *79*, 101–106. [CrossRef]
42. Podwyszyńska, M.; Cieślińska, M. Rooting shoots of apple varieties and their tetraploids obtained by the In Vitro technique. *Acta Sci. Pol. Hortorum Cultus* **2018**, *17*, 49–62. [CrossRef]
43. Podwyszyńska, M.; Pluta, S. In vitro tetraploid induction of the blackcurrant (*Ribes nigrum* L.) and preliminary phenotypic observations. *Zemdirb.-Agric.* **2019**, *106*, 151–158. [CrossRef]
44. Clark, M.F.; Adams, A.N. Characteristics of the microplate method of enzyme-linked immunosorbent assay for the detection of plant viruses. *J. Gen. Virol.* **1977**, *34*, 475–483. [CrossRef] [PubMed]
45. Murashige, T.; Skoog, F. A revised medium for rapid growth and bioassays with tobacco tissue cultures. *Physiol. Plant.* **1962**, *15*, 473–497. [CrossRef]
46. Nielsen, I.L.; Haren, G.R.; Magnussen, E.L.; Dragsted, L.O.; Rasmussen, S.E. Quantification of anthocyanins in commercial black currant juices by simple high-performance liquid chromatography. Investigation of their pH stability and antioxidative potency. *J. Agric. Food Chem.* **2003**, *51*, 5861–5866. [CrossRef] [PubMed]
47. Dogan, Y.; Baslar, S.; Ay, G.; Mert, H.H. The use of wild edible plants in western and central Anatolia (Turkey). *Econ. Bot.* **2004**, *58*, 684–690. [CrossRef]
48. Penafiel, D.; Vanhove, W.; Espinel, R.L.; Van Damme, P. Food biodiversity includes both locally cultivated and wild food species in Guasaganda, Central Ecuador. *J. Ethn. Foods* **2019**, *6*, 25. [CrossRef]
49. Perrino, E.V.; Valerio, F.; Gannouchi, A.; Trani, A.; Mezzapesa, G. Ecological and Plant Community Implication on Essential Oils Composition in Useful Wild Officinal Species: A Pilot Case Study in Apulia (Italy). *Plants* **2021**, *10*, 574. [CrossRef] [PubMed]

plants

MDPI

Article

Repression of Carotenoid Accumulation by Nitrogen and NH_4^+ Supply in Carrot Callus Cells In Vitro

Tomasz Oleszkiewicz *, Michał Kruczek and Rafal Baranski

Department of Plant Biology and Biotechnology, Faculty of Biotechnology and Horticulture, University of Agriculture in Krakow, 31-425 Krakow, Poland; kruczek.michael@gmail.com (M.K.); rafal.baranski@urk.edu.pl (R.B.)
* Correspondence: tomasz.oleszkiewicz@urk.edu.pl

Abstract: The effect of mineral nutrition on the accumulation of the main health beneficial compounds in carrots, the carotenoid pigments, remains ambiguous; here, a model-based approach was applied to reveal which compounds are responsible for the variation in carotenoid content in carrot cells in vitro. For this purpose, carotenoid-rich callus was cultured on either BI (modified Gamborg B5) or R (modified Murashige and Skoog MS) mineral media or on modified media obtained by exchanging compounds between BI and R. Callus growing on the BI medium had abundant carotene crystals in the cells and a dark orange color in contrast to pale orange callus with sparse crystals on the R medium. The carotenoid content, determined by HPLC and spectrophotometrically after two months of culture, was 5.3 higher on the BI medium. The replacement of media components revealed that only the N concentration and the NO_3:NH_4 ratio affected carotenoid accumulation. Either the increase of N amount above 27 mM or decrease of NO_3:NH_4 ratio below 12 resulted in the repression of carotenoid accumulation. An adverse effect of the increased NH_4^+ level on callus growth was additionally found. Somatic embryos were formed regardless of the level of N supplied. Changes to other media components, i.e., macroelements other than N, microelements, vitamins, growth regulators, and sucrose had no effect on callus growth and carotenoid accumulation. The results obtained from this model system expand the range of factors, such as N availability, composition of N salts, and ratio of nitrate to ammonium N form, that may affect the regulation of carotenoid metabolism.

Keywords: *Daucus carota*; carotene; nitrate; ammonium; somatic embryogenesis

Citation: Oleszkiewicz, T.; Kruczek, M.; Baranski, R. Repression of Carotenoid Accumulation by Nitrogen and NH_4^+ Supply in Carrot Callus Cells In Vitro. *Plants* **2021**, *10*, 1813. https://doi.org/10.3390/plants10091813

Academic Editors: Laura Pistelli and Kalina Danova

Received: 10 August 2021
Accepted: 27 August 2021
Published: 31 August 2021

1. Introduction

The carrot is a well-known vegetable grown around the world for its nutritious storage root. The roots of the most common carrot varieties accumulate carotenoids, mainly β-carotene and α-carotene, which give them their orange color. Both carotenes have provitamin A activity and, together with other carotenoids, play beneficial roles in human health [1]. The high carotene content makes carrots one of the most important sources of carotenoids in the human diet [2], and knowledge on carotenoid biosynthesis, accumulation, and regulation of these processes is essential for the development of high-quality carrot varieties.

Carotenoids exist widely in nature. They are 40-carbon molecules built from eight base isoprenoid units. They are classified to two main groups: carotenes, being hydrocarbons such as β-carotene and α-carotene, and xanthophylls, which are oxidized carotenes [3]. The processes of carotenoid biosynthesis in plants, including carrot, have been well described [4]. In recent years, research has focused on understanding the regulation of biosynthetic pathway and carotenoid sequestration. Currently, it is known that developmental and environmental factors, such as light, influence carotenoid accumulation in carrot cells [5,6]. Field conditions and genotype have a pronounced effect on carotenoid accumulation in carrot storage roots in contrast to plant fertilization [7]. However, fertilization with calcium ammonium nitrate increased carotenoid content [8]. The effect of other N salts was equivocal unless an additional foliar nutrition with a complex fertilizer was applied [9]. Variation in carotenoid content among carrot cultivars was also reported depending on the applied urea dose [10]. Thus, the conclusions regarding the effect of nutrition on carotenoid metabolism in carrot remain ambiguous. Recently, it was demonstrated that ammonium ions negatively affect carotenoid accumulation in *Calendula officinalis* callus cultured in vitro [11]. Another genetic research based on callus response to changes in the composition of mineral medium indicated that N supply affected pigment accumulation [12].

For more than 60 years, the carrot has been considered a model species in research on totipotency, somatic embryogenesis, and horizontal gene transfer (for review see [13] and [14]), while broad genetic research, including the recent genome sequencing project [15], led to the development of high-quality varieties [16]. Carrot is amenable to cell and tissue culture on mineral media in vitro. Callus can be induced in in vitro culture from various explants by the supplementation of mineral medium with growth regulators and then it can be easily propagated [17], hence, it has become a convenient material for research on stress factors, genetic transformation, and genome editing, including genes of the carotenoid pathway [18,19]. Carrot callus usually accumulates low amounts of carotenoids, as do the storage root meristematic cells used to induce callus [20,21]. However, the development of carotenoid-rich callus was also reported [22–24], and recently, it has been successfully used for structural studies of carotene crystals [25,26], regulation of carotenoid biosynthesis, sequestration, and interaction of carotenoid and cell wall composition when it was subjected to targeted mutagenesis using novel tools of genome editing, i.e., clustered regularly interspaced short palindromic repeats (CRISPR) and CRISPR associated (Cas9) proteins [19,24].

Inorganic components in the medium are necessary for the growth of plant tissues and organs in vitro. Their contents and composition influence tissue and plantlet development, hence, the proper balance of medium components is required [27]. For the induction of carrot callus development, and for its further propagation, a medium based on Gamborg B5 [28] mineral salts and vitamin composition is more effective and more often used than the Murashige and Skoog (MS; [29]) medium [30]. The MS medium is recommended for somatic embryogenesis and carrot plant regeneration [31,32]. Our preliminary experiments showed that attempts of plant regeneration using a modified MS medium, the R medium, led to a visually paler color of callus due to a reduced carotenoid accumulation in comparison to callus grown on the B5-based BI medium. Both media, BI and R, differ in the composition of salts, vitamins, plant growth regulators, and in the sucrose concentration. Thus, in this study we sought for an answer to which component of the R medium is responsible for the repression of carotenoid accumulation. For this purpose, we used a model carotenoid-rich callus [23] and exposed it to media with modified compositions. A substantial reduction of carotenoid content was observed on the B5 medium with altered composition of nitrogen salts, thus, we show here that the amount of N and NO_3:NH_4 ratio are key factors affecting carotenoid accumulation in carrot cells.

2. Results

2.1. Callus Growth and Morphology

Carrot callus cultured on both BI (modified Gamborg B5) and R (modified MS) media (Supplementary Table S1) grew with a similar rate. Callus cultured on the BI medium retained its characteristic morphology throughout all experiments. It had dense and lumpy structure, with small parts being more friable. It retained orange color, although rarely, friable callus was paler (Figure 1a). Callus growing on the R medium differed from callus on the BI medium in color, which became paler and eventually light orange over the course of time (Figure 1b). Microscopic observations revealed that regardless of the medium, callus cells were densely packed in aggregates (Figure 2a,b). Carotene crystals, clearly distinguishable due to their intense orange color, were sequestered in the cells of dark orange callus on the BI medium (Figure 2c). Cells of callus maintained on the R medium contained only small crystals, and they were not abundant (Figure 2d). Additionally, proembryogenic tissue was identified in callus on the R medium. Embryoid structures were visible (Figure 1b inset); however, their development was arrested at early stages. They did not convert into plants, turned brown, and often died or dedifferentiated to new callus cell layers.

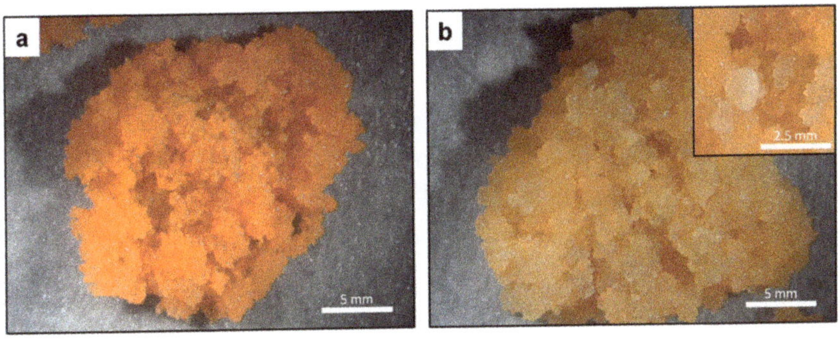

Figure 1. Carrot callus growing on mineral media. (**a**) BI medium; (**b**) R medium; (**b-inset**) embryo-like structure.

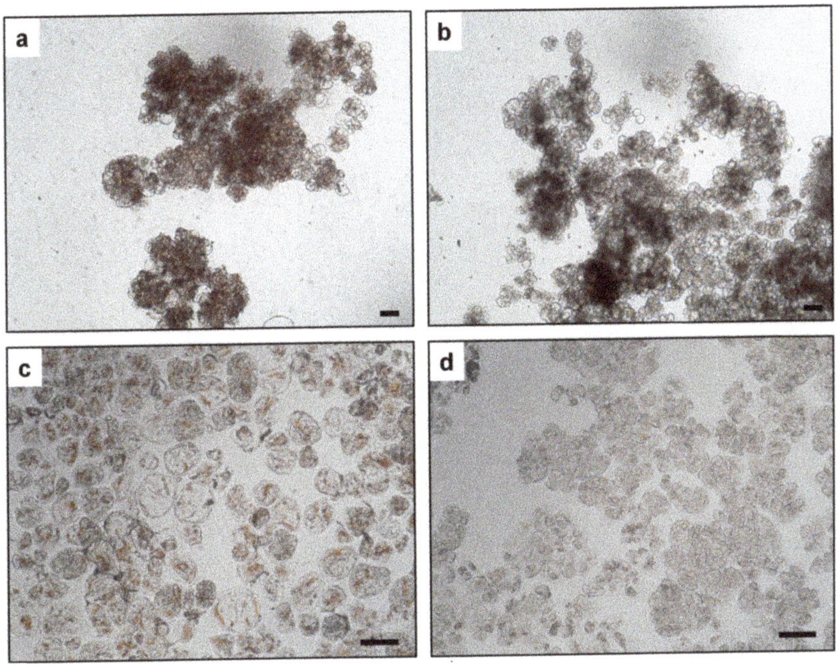

Figure 2. Densely packed carrot callus cells (not macerated tissue) growing on either the BI (**a**) or R medium (**b**). Easily noticeable carotene crystals in cells growing on the BI medium (**c**) and sparse crystals in cells growing on the R medium (**d**). Bar = 100 μm.

2.2. Carotenoid Content in Callus

Two main carotenes were identified in callus using HPLC (Table 1 and Supplementary Figure S1). The sum of α- and β-carotene contents in callus growing on the BI medium was very high (2264 μg/g DW) and exceeded that in callus on the R medium (425 μg/g DW) by 5.3 times, which corresponded to differences in color observed between calli on both media. The β/α carotene ratio (2.6) was also higher for the BI medium than for the R medium (1.5) (Table 1). The sum of α- and β-carotenes determined by HPLC highly correlated ($r = 0.98$, $p < 0.001$) with the total carotenoid content, determined spectrophotometrically, although it was usually higher. A linear relationship was described by a well fitted regression line ($p < 0.001$) with the coefficient of determination $R^2 = 0.96$ (Figure 3). Hence, quantitative determination of carotenoid content in further experiments was done using spectrophotometry.

Table 1. Carotenoid content [μg/g DW] in callus growing on the BI and R media.

Medium	α-carotene [1]	β-carotene [1]	β:α Ratio	Total Carotenoids [2]
BI (modified Gamborg B5 medium)	627 ± 73 a	1637 ± 305 a	2.6	1169 ± 89 a
R (modified MS medium)	172 ± 55 b	253 ± 113 b	1.5	404 ± 62 b
BI/MS-macro (BI with macroelements as in MS (R))	304 ± 35 b	529 ± 101 b	1.7	531 ± 44 b
R/B5-macro (R with macroelements as in B5 (BI))	725 ± 55 a	2191 ± 182 a	3.0	1322 ± 99 a
BI:R ratio	3.6	6.5	nd [3]	2.9

[1] determined by HPLC, [2] determined spectrophotometrically; means ± std. error ($n = 4$); means followed by different letters within column are significantly different at $p = 0.05$; [3] not determined.

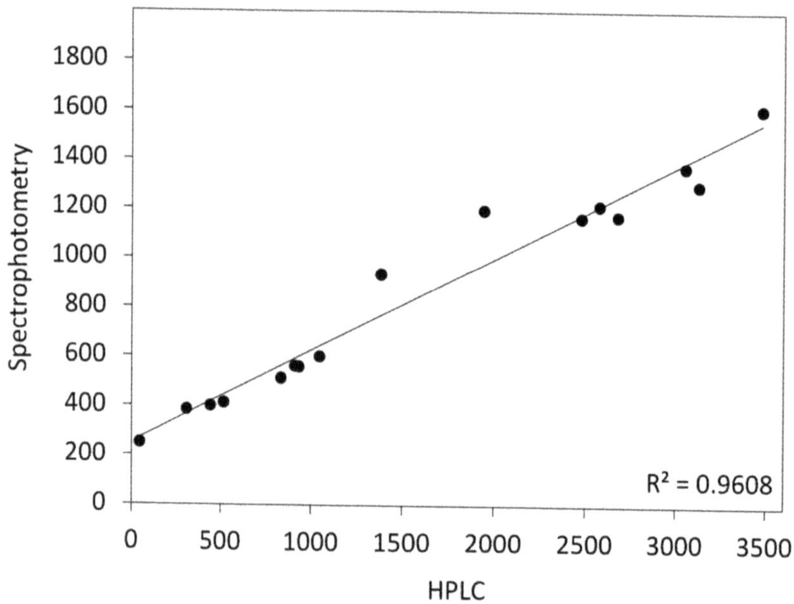

Figure 3. Fitted linear regression model for carotenoid content (μg/g DW) in carrot callus, determined based on HPLC (sum of α- and β-carotene) and spectrophotometric measurements (total carotenoids).

2.3. Effect of Medium Composition

To identify compounds that affected the carotenoid content in callus, 12 media varying in the composition of main compound groups were compared. For this purpose, the composition and amounts of macroelements, microelements, vitamins, growth regulators, or sucrose in the BI medium were replaced by the corresponding compound groups, and in the same amounts, as present in the R medium (Table 2). Analogous modifications were applied to the R medium by replacing compound groups to be the same as in the BI medium. Callus growing on the modified media showed changes in color and carotene contents ($p < 0.001$). The replacement of macroelements in the BI medium caused callus discoloration and the decrease of carotenoid content by 54.6% to a level similar to the R medium (Table 2). The effect of B5 macroelements added to the R medium was also significant. Callus growing on the R/B5-macro medium had an intense orange color and over 3-fold increased carotenoid content in comparison to the unmodified R medium, reaching the carotenoid level present in callus on the BI medium (Table 2). Thus, the Gamborg B5 composition of macroelements stimulated carotenoid accumulation in contrast to the MS formulation of macroelements. Any other changes done to either the BI or R media composition (microelements, vitamins, growth regulators, sucrose) did not significantly affect carotenoid content (Table 2). These results indicated that macroelement composition in the medium was critical for carotenoid accumulation in callus. The only other effect of modified media was the formation of embryoid structures on the BI medium free of 2,4-D and kinetin that resembled structures observed in callus on the R medium.

Table 2. Modified media used for callus culture and the carotenoid content in callus after eight-week culture (mean ± std. error).

Experiment	Medium	Medium Modification	Carotenoid Content (µg/g DW)	%BI [1]	*P* (BI) [2]	*P* (R) [2]
Modification of compound groups	BI [3]	Modified Gamborg B5 medium	1169 ± 89	100.0	ref	*
	BI/MS-macro	BI with macroelements as in MS (R)	531 ± 44	45.4	*	ns
	BI/MS-micro	BI with microelements as in MS (R)	1292 ± 64	110.5	ns	*
	BI/MS-vit	BI with vitamins as in MS (R)	1226 ± 80	104.8	ns	*
	BI/MS-pgr	BI without growth regulators as in R	1279 ± 91	109.4	ns	*
	BI/MS-suc	BI with 2% sucrose as in R	1386 ± 225	118.5	ns	*
	R [4]	Modified MS medium	404 ± 62	34.5	*	ref
	R/B5-macro	R with macroelements as in B5 (BI)	1322 ± 99	113.0	ns	*
	R/B5-micro	R with microelements as in B5 (BI)	554 ± 44	47.4	*	ns
	R/B5-vit	R with vitamins as in B5 (BI)	485 ± 51	41.5	*	ns
	R/B5-pgr	R with growth regulators as in BI	505 ± 56	43.2	*	ns
	R/B5-suc	R with 3% sucrose as in BI	512 ± 99	43.8	*	ns

Table 2. *Cont.*

Experiment	Medium	Medium Modification	Carotenoid Content (μg/g DW)	%BI [1]	*P* (BI) [2]	*P* (R) [2]
Modification of macro-elements	BI	as above	1210 ± 71	100.0	ref	*
	R	as above	656 ± 52	54.3	*	ref
	R+K	R suppl. with 2.35 mM K_2SO_2	736 ± 128	60.8	*	ns
	BI/MS-N	BI with nitrogen salts as in MS (R)	753 ± 178	62.2	*	ns
	BI/MS-N+K	BI with nitrogen salts as in MS (R) suppl. with 2.97 mM K_2SO_2	736 ± 83	60.8	*	ns
	BI/MS-P	BI with phosphorus salts as in MS (R)	1216 ± 202	100.5	ns	*
	BI/MS-Mg	BI with magnesium salts as in MS (R)	1300 ± 162	107.5	ns	*
	BI/MS-Ca	BI with calcium salts as in MS (R)	1435 ± 107	118.6	ns	*

[1] %BI—carotenoid content expressed as the percentage of the content in callus growing on the BI medium; [2] *P*—significant at $p < 0.05$ (*) or not significant (ns) difference from either BI or R considered as the reference (ref) according to the Dunnett test; [3] Modified Gamborg B5 [28] medium; [4] Modified MS [29] medium.

2.4. Effect of Macroelements

Various salts of macroelements were present in the BI and R media, thus, any element or their combination could affect carotene content. The contents of individual macroelements in the BI medium were modified to get the same molar concentrations as in the R medium. Modifications to either P, K, Ca, or Mg contents did not result in changes either of callus morphology, color, or carotenoid content. A noticeable callus discoloration was observed only on the medium with a modified composition of N salts. Callus exposed to the BI/MS-N medium developed more white or pale orange cell aggregates. It accumulated almost 40% less carotenoids than callus on the BI medium and had similar amounts of carotenoids as callus on the R medium (Table 2). The N content depends on the amounts of NH_4NO_3 and KNO_3 salts, and KNO_3 supplementation results in the increased concentration of both N and K. To verify the effect of K on carotenoid content, the R and BI/MS-N media were supplemented with K_2SO_4 (R+K and BI/MS-N+K, respectively). No effect of additional K amounts on carotenoid level was found.

2.5. Effect of N Concentration and NO_3:NH_4 Ratio

The N content in the R medium (60.02 mM) was more than doubled in comparison to the BI medium (26.76 mM) (Table 3). To verify the effect of N concentration on carotenoid accumulation, media differing in composition of N salts were compared (Supplementary Table S2). The increase of N amount in the range from 27 mM to 80 mM did not affect callus growth but significantly reduced carotene content from 1252 μg/g DW to 411 μg/g DW. Such changes were highly significant independent from whether the NO_3:NH_4 ratio in all comparing media was the same, i.e., 12.19 (the same as in the BI medium) or it was increasing in the range from 12.19 up to 38.5 (both $p < 0.001$). In both sets of media, the observed reduction of carotene content followed similar trends described by logarithmic functions with R^2 of 0.8249 and 0.9490, respectively (Figure 4).

The R medium contained NH_4NO_3 not present in the BI medium, which had NH_4^+ ions supplied in a low amount of $(NH_4)_2SO_4$ not present in the R medium (Table 3). In consequence, the R medium had more N, mainly due to the use of an amount 10.2 times higher of the ammonium form while the nitrate form was only 1.6 times higher. The effect of N form in the medium on the carotene content was verified by using media containing 26.76 mM N, the same as in the BI medium, adjusted by using both nitrate and ammonium salts in various ratios. While decreasing the NO_3:NH_4 ratio down to 1.91 (the same as in the

R medium), callus grew similar to the callus exposed to the BI medium with the NO_3:NH_4 ratio of 12.19. Further elevation of the NH_4^+ amount in the media restricted callus growth, which was eventually inhibited on the medium with the 1:1 NO_3:NH_4 ratio. The increasing NH_4^+ level also highly reduced carotenoid content in callus ($p < 0.001$). In comparison to the BI medium, the carotenoid content was reduced 3.0-fold at the NO_3:NH_4 ratio of 1.91 (the same as in the R medium), and a further increase of NH_4^+ to the 1:1 NO_3:NH_4 ratio reduced the carotenoid content by 11.5-fold to the level of 115 μg/g DW. Such changes in the carotenoid content followed a trend described by a well fitted logarithmic function with $R^2 = 0.9685$ (Figure 5).

Table 3. The BI and R media composition with regard to nitrogen content.

N Form	Compound/Ion/N	BI (mM)	R (mM)	BI:R Ratio
Salt	KNO_3	24.73	18.79	1.3
	$(NH_4)_2SO_4$	1.01	0	nd [1]
	NH_4NO_3	0	20.61	nd
Ion	NO_3^-	24.73	39.41	0.6
	NH_4^+	2.02	20.61	0.1
Element	N	26.76	60.02	0.4
Ratio	NO_3:NH_4	12.19	1.91	6.4

[1] not determined.

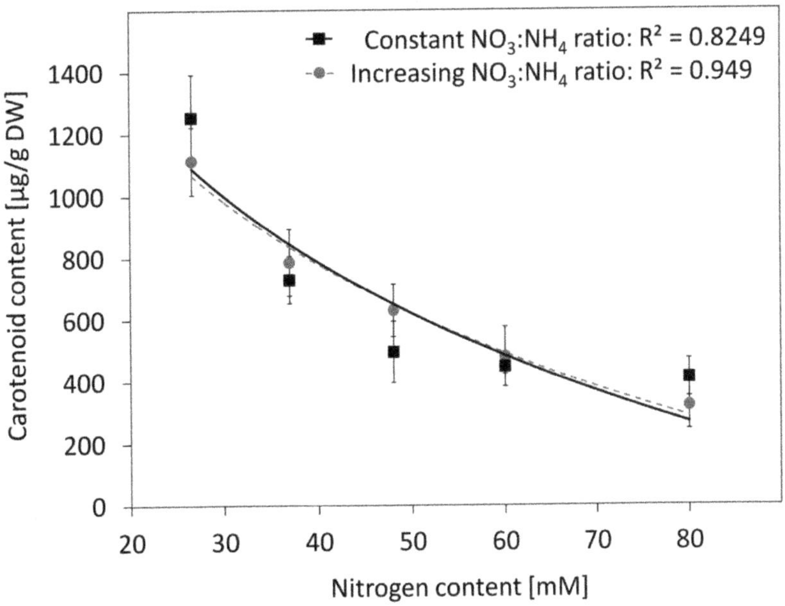

Figure 4. Carotenoid content in callus grown on the BI media with modified N content and with the constant (12.2) NO_3:NH_4 ratio (squares, solid line) or with increasing NO_3:NH_4 ratio from 12.2 to 38.5 (dots, dashed line). Lines represent logarithmic functions: $y = 3551.5 - 749.5\ln(x)$ for the constant NO_3:NH_4 ratio and $y = 3852 - 820.7\ln(x)$ for the increasing NO_3:NH_4 ratio; whiskers—std. error.

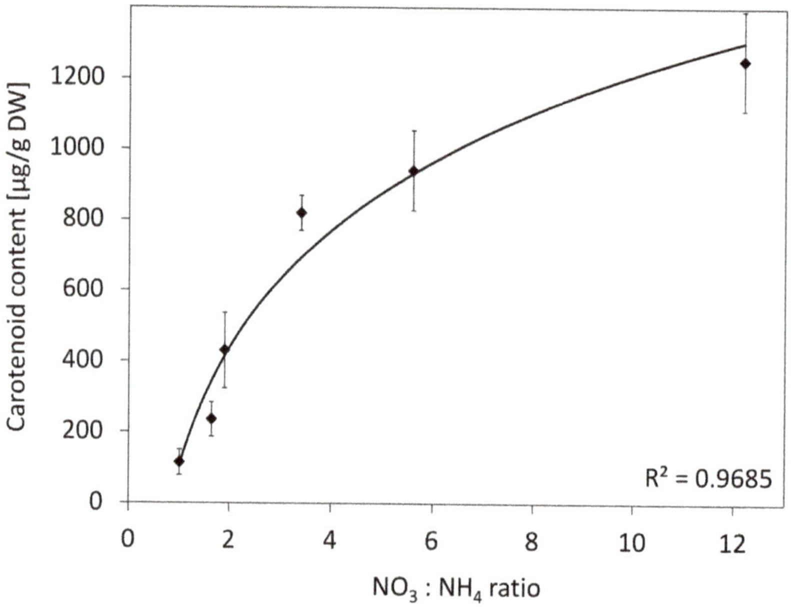

Figure 5. Carotenoid content in callus grown on BI media with a modified NO_3:NH_4 ratio and with the constant N content (26.8 mM). The line represents a logarithmic function y = 107.65 + 478.27ln(x); whiskers—std. error.

3. Discussion

Nutrient supply and their uptake by plants determine yield and quality of agricultural products, including carrot, and available data indicate that fertilization, in particular with N, may also affect carotenoid accumulation in carrot storage roots. Previous evaluations of NPK fertilization showed that genotype and environmental conditions affected carotenoid accumulation rather than N supply [7]. These conclusions were supported by results of a multiyear field trial using various N fertilizers, although significant increase of carotenoid content was achieved after foliar nutrition [9]. Fertilization with urea suggested variation in carotenoid content in used cultivars depending on the urea dose, although differences between overall means were insignificant [10], while fertilization with calcium ammonium nitrate increased carotenoid accumulation in two cultivars in a two-year trial [8]. The conclusions of field studies on plant nutrition remain ambiguous as the results are highly affected by complex environmental factors, additionally interacting with variety.

Experiments utilizing cell and tissue culture in vitro allow to apply controlled conditions that, in particular, are essential in plant nutrition research, and which are not possible to obtain in field conditions. Therefore, we have applied a research model to elucidate at cellular level the role of nutrition on the accumulation of the main carrot health beneficial compounds, the carotenoid pigments. The MS-based mineral media had already been used to culture cell suspension or to induce callus for carotenoid research. The cell suspension or callus from a red storage root carrot variety accumulated mainly β-carotene and lycopene; the level of these pigments highly varied and was clone-dependent [33,34]. For *Arabidopsis thaliana*, the carotenoid content in wild type callus cultured on the medium containing MS salts was low, 200–550 µg/g DW [35]. For *Tagetes erecta* [36], individual carotenoids were identified and not quantified, but the assessment of color and HPLC profiles also indicated low amounts of pigments. The Gamborg B5-based mineral media were used to induce development of light-orange [24] and dark-orange, carotenoid-rich, carrot callus accumulating up to the same amounts of carotenoids (2150 µg/g DW) as the storage root from which such callus was derived [23]. Thus, carotenogenesis was ongoing in materials cultured on the MS-based media, but the B5-based media were much more efficient for pigment accumulation. The observed color variation of carrot callus cultured on different mineral media in vitro have indicated that accumulation of carotenoid pigments is stimulated or repressed by media components. The mineral compositions of BI and R media, used in this work, differed significantly as they were essentially based on the Gamborg B5 and MS formulations, respectively. Both media differed mainly in their N salts composition. The amount of N was 2.24 times higher in the MS medium, and N was supplied in NH_4NO_3 and KNO_3 salts, of which the former was present in a higher concentration, thus, the NO_3:NH_4 ratio in MS was 1.91 (Table 3). The B5 medium was richer in the nitrate salt by 32% but contained a low amount of ammonium $(NH_4)_2SO_4$ salt, thus, the NO_3:NH_4 ratio in B5 was 12.19. Hence, the B5 medium contained 10.2 times less ammonium N form than MS. In this study, we found that callus grown on the BI medium and on any modified BI medium containing N salts according to the Gamborg B5 formulation accumulated many more carotenoids than when using MS-based N salts.

Both BI and R media differed also in the composition of other compounds. The MS medium contained three times more $CaCl_2$, 50% more $MgSO_4$, and had KH_2PO_4 instead of NaH_2PO_4. Although these differences are less pronounced than differences in N salts composition, they were also taken into account in this study. It was previously reported that a reduction of Ca supply promoted carotenoid accumulation in the roots of carrot plants, but this effect was variety dependent, with the most significant effect on lycopene content in a lycopene accumulating variety [37]. When using a lycopene accumulating carrot cell suspension, it was shown that increasing the initial P content in the medium or resupplying P during the culture increased carotenoid accumulation [38]. It was also shown that the increase of 2,4-D up to 10 ppm promoted carotenoid accumulation in carrot cells [33]. A higher sucrose concentration increased carotenoid content, with 3%–5% sucrose being optimal, while 8% sucrose had adverse effects on carrot cells growth and their size [39]. Additionally, the carotenoid accumulation increased in *Calendula officinalis* callus when sucrose concentration was raised from 4% to 7% [11]. In our work, no significant changes in carotenoid accumulation in carrot callus was found when modifying the compositions of Ca, P, K, and Mg salts. Further BI medium modifications by replacing microelements, vitamins, elimination of growth regulators, and reduction of sucrose from 3% to 2%, as present in the R medium, had no significant impact. No response of carrot callus to all these modifications supports the conclusion that N availability is the prime factor affecting carotene accumulation. This finding is in contrary to the results presented by Hanchinal et al. [40], who modified N, P, and sucrose concentrations and used a response surface methodology to optimize β-carotene production by carrot cells in suspension. A doubled N concentration to 50 mM with increased sucrose content from 2% to 3% increased β-carotene production up to 13.61 μg/g DW. However, it must be underlined that they used cells accumulating very low amounts of carotenoids, two magnitude lower than callus in our work, which may highly bias the conclusions.

Our results showed also that a gradual increase of N from 26.76 mM to 80.04 mM restricts carotenoid accumulation, which eventually decreased 3-fold. Nitrogen was supplied mainly in the form of KNO_3, thus, the amounts of N and K in the medium were interrelated. Further media adjustments with K salts to keep this element at the same level while increasing N concentration showed that K did not affect carotenoid accumulation, confirming that the N amount in the medium is a critical factor and, moreover, its effect is independent on the NO_3:NH_4 ratio in the range from 12.2 to 38.5. Additionally, the amount of N did not alter callus growth. Recent study on grape callus showed that the reduced N amount in the MS medium from 60 mM to 40–50 mM enhanced accumulation of other pigments, anthocyanins, in red-pod okra callus; however, further reduction to 30 mM had an adverse effect [41]. Additionally, N starvation promoted anthocyanin accumulation in grape callus [42]. In contrary, other reports showed that the increase of total N content by doubling KNO_3 in the MS medium increased anthocyanin content by 135% [12], which was congruent with results showing the highest accumulation of anthocyanins using an elevated N amount (70 mM) in the medium [43].

The N content in a medium depends on the combination of supplied ammonium and nitrate salts. The ammonium N form is preferred by plants as it can be directly used, and its incorporation by a cell requires less energy. However, it can become toxic to plant cells at higher concentrations, and plant sensitivity to ammonia varies greatly depending on species, plant age, and environment pH [44]. Hairy roots of carrot, red beet, and madder in the presence of NH_4^+ available in the amounts in the MS medium had a reduced growth [45]. A similar effect of restricted growth was observed for anthocyanin accumulating carrot callus cultured on MS [12]. No growth changes were reported only when carrot cell suspension was exposed to a doubled amount of ammonium N form than present in MS [46]. The comparison of a wide range of NO_3:NH_4 ratios in our work demonstrates that when keeping the optimum N level (26.76 mM) for carotenoid accumulation, as in the B5 medium, the callus growth is restricted with increasing amounts of NH_4^+, as is the carotenoid content. This adverse effect of the ammonium form led to an over 10-fold reduction of carotenoids and such response intensified logarithmically with the NH_4^+ concentration. A similar adverse effect of high NH_4^+ concentration was found in *Calendula officinalis* callus. The 50% decrease of NH_4^+ amount, in comparison to MS, induced carotenoid biosynthesis, and the complete removal of NH_4^+ from the medium further promoted carotenoid accumulation [11]. An analogous effect of increasing NH_4^+ concentration was reported in relation to anthocyanin accumulation. Lower anthocyanin contents were recorded in carrot cells in suspension exposed to media with a low NO_3:NH_4 ratio, and the increase of the ratio to 4:1 led to the highest pigment content [43]. In callus induced from rose leaves, the reduction of NH_4^+ and increase of NO_3^- concentrations in the MS-based medium enhanced anthocyanin accumulation [47]. Conversely, doubling the concentration of NH_4^+ in the MS medium restricted the anthocyanin content by one third in carrot callus [12].

Nitrogen is required by plants for their growth, but the composition and concentrations of N salts in a culture medium affect also morphogenesis and embryogenesis [27]. It was previously shown that a reduced form of N, present in high amounts in MS, is required for the development of somatic embryos from carrot hypocotyl explants. Three N salts with reduced N were compared and two of them, NH_4NO_3 and NH_4Cl, favored somatic embryogenesis, while $(NH_4)_2SO_4$ did not [48]. However, contrary to these results, nearly 20 times more plants developed on the B5 medium, which contained much less of the reduced N form and was supplied with $(NH_4)_2SO_4$ salt only, than on the MS-based medium [49]. The comparison of several media free of growth regulators in our work has shown that the formation of proembryogenic tissue and globular embryos is ongoing regardless of the medium mineral composition. Somatic embryos were observed on both, carotenoid-rich callus and low carotenoid accumulating callus, thus, there was no clear relationship between somatic embryogenesis and N salt composition, hence, carotenoid accumulation, although quantitative comparison was not the subject of this work.

4. Materials and Methods

4.1. Plant Material, Media Preparation, and Experiment Design

Dark-orange callus derived from the root of DH1 (doubled haploid) carrot (*Daucus carota* L.) line accumulating high amounts of carotenoids and described previously [23] was used. Callus was maintained on filter paper disks laid down on the surface of solidified mineral medium in 9 cm Petri dishes and cultured at 26 °C in the dark. In each experiment, callus was grown for eight weeks with one transfer to a fresh medium after four weeks. Two main media were used: (1) the BI medium consisting of Gamborg B5 macro- and microelements with vitamins, 30 g/L sucrose, 1 mg/L 2,4-D, and 0.0215 mg/L kinetin, and (2) the R medium consisting of MS macro- and microelements with vitamins (with an increased glycine content to 3 mg/L), 20 g/L sucrose, and free of growth regulators (Supplementary Table S1). Both media had pH adjusted to 5.8 and were solidified with

2.7 g/L phytagel. Macro- and microelements, including vitamin mixtures, separate macro- and microelement mixtures, vitamins, and plant growth regulators were purchased from Duchefa Biochemie (Haarlem, The Netherlands). Media modifications were done by exchanging group of components between BI and R media (12 media variants; Table 2), or by exchanging individual compounds (eight variants; Table 2), or by changing nitrogen salts composition and their concentration (14 variants) (Supplementary Table S2).

4.2. Microscopic Observations

To observe callus structure, small callus pieces were placed on a microscopic slide in a water drop under the cover slide. Carotenoid crystals were observed in single cells after tissue maceration in 1N HCl at 50 °C for 5 min. Observations were done in a bright-field using the Zeiss Axiovert S100 microscope with ×10 objective. Images were collected by using the attached digital camera.

4.3. Determination of Carotenoid Content

Eight-week-old callus was lyophilized and ground into a fine powder in a beading mill for 5 min. Carotenoids were extracted from 5–10 mg samples with 500 µL of acetone in 1.5 mL tubes. Samples were vortexed for 30 s and centrifuged at 18,000 g for 5 min. Acquired extracts were transferred to fresh tubes. The procedure was repeated to ensure complete extraction, and then, obtained extracts were combined. The absorbance of extracts was measured in a 1 cm QS quartz cuvette (Hellma Analytics, Müllheim, Germany) at 450 nm using the NanoDrop 2000c (ThermoScientific, Waltham, MA, USA) spectrophotometer. Extractions and measurements were done for each sample in triplicate and the readouts were averaged before statistical analysis. The total carotenoid content was calculated based on the β-carotene extinction coefficient ($A_{1cm}^{1\%}$ = 2500) using the formula:

$$\frac{\frac{absorbance\ (450\ nm)}{2500} \times extract\ volume\ (ml) \times 10000}{sample\ mass\ (g)}$$

The results are presented in µg of carotenes per gram of callus dry weight.

High performance liquid chromatography (HPLC) measurements were done using the same callus samples as for spectrophotometry. HPLC was performed as previously described [19]. Briefly, the extraction was performed using ethanol:*n*-hexane (1:1, *v:v*) and HPLC was performed using Shimadzu LC–20AD chromatograph equipped with a C18 RP (5 µm) column and the Shimadzu SPDM–20A–DAD photodiode-array detector. The identification of β-carotene was based on the retention time of the standard and confirmed by analysis of absorption spectra. The identification of α-carotene was based on the analysis of the absorption spectra. Quantification of β-carotene was done using a standard curve, while α-carotene was quantified in relation to β-carotene.

4.4. Statistical Analysis

Each experiment was set up in four replicates, each consisting of five calli, and having a completely randomized design. A one-way ANOVA was performed to test effects of media composition on carotene contents in callus using the Statistica v.13.1 software (TIBCO; Palo Alto, CA, USA). Differences between means were verified at the significance level $p = 0.05$ using the Dunnett test. Means are presented with their standard errors.

5. Conclusions

In this study, we sought for the answer to which component of the culture medium affects carotenoid accumulation in carrot callus. A compound by compound replacement in the MS and Gamborg B5 media have revealed that the only critical element is nitrogen, and either the increase of the total N concentration or the decrease of NO_3:NH_4 ratio restricts carotenoid accumulation. Thus, the highest carotenoid content was achieved in the medium with 26.76 mM N and 12.19:1 NO_3:NH_4 ratio, regardless of whether the other media components were supplied according to the MS or Gamborg B5 formulation. These model-based obtained results pave the way for further elucidation of biological processes related to regulation of carotenoid metabolism. The observed effects might be limited to a simplified in vitro model; hence, further confirmation in planta may be required. However, they can be useful for research or application purposes using cell or tissue culture where stimulation of valuable secondary metabolites, such as carotenoids, is required.

Supplementary Materials: The following are available online at https://www.mdpi.com/article/10.3390/plants10091813/s1, Table S1: Composition of Gamborg B5 and Murashige and Skoog (MS) media and their modified variants, BI and R, respectively; Table S2: Nitrogen salts composition of BI media with a modified N content or NO_3:NH_4 ratio; Figure S1: HPLC chromatogram at 452 nm of the sample from callus grown on a BI (control) medium.

Author Contributions: Conceptualization, T.O.; methodology, T.O. and M.K.; formal analysis, T.O. and R.B.; investigation, T.O. and M.K.; resources, T.O.; data curation, T.O.; writing—original draft preparation, T.O.; writing—review and editing, R.B.; visualization, T.O.; supervision, R.B.; project administration, T.O.; funding acquisition, T.O. All authors have read and agreed to the published version of the manuscript.

Funding: This Research was financed by the Ministry of Science and Higher Education of the Republic of Poland. This research was supported by the National Science Centre, Poland (grant No. 2018/31/N/NZ9/02368).

Institutional Review Board Statement: Not applicable.

Informed Consent Statement: Not applicable.

Data Availability Statement: Not applicable.

Acknowledgments: Tomasz Oleszkiewicz received financial funding for a doctoral scholarship from the National Science Centre, Poland (Etiuda No. 2019/32/T/NZ9/00463).

Conflicts of Interest: The authors declare no conflict of interest.

References

1. Eggersdorfer, M.; Wyss, A. Carotenoids in human nutrition and health. *Arch. Biochem. Biophys.* **2018**, *652*, 18–28. [CrossRef]
2. Simon, P.W. Carrots and Other Horticultural Crops as a Source of Provitamin A Carotenes. *HortScience* **1990**, *25*, 1495–1499. [CrossRef]
3. Rodriguez-Amaya, D.B. Structures and Analysis of Carotenoid Molecules. In *Carotenoids in Nature: Biosynthesis, Regulation and Function*; Stange, C., Ed.; Springer International Publishing: Cham, Switzerland, 2016; pp. 71–108. [CrossRef]
4. Rosas-Saavedra, C.; Stange, C. Biosynthesis of carotenoids in plants: Enzymes and color. In *Carotenoids in Nature: Biosynthesis, Regulation and Function*; Stange, C., Ed.; Springer International Publishing: Cham, Switzerland, 2016; pp. 35–69. [CrossRef]
5. Rodriguez-Concepcion, M.; Stange, C. Biosynthesis of carotenoids in carrot: An underground story comes to light. *Arch. Biochem. Biophys.* **2013**, *539*, 110–116. [CrossRef]
6. Perrin, F.; Hartmann, L.; Dubois-Laurent, C.; Welsch, R.; Huet, S.; Hamama, L.; Briard, M.; Peltier, D.; Gagné, S.; Geoffriau, E. Carotenoid gene expression explains the difference of carotenoid accumulation in carrot root tissues. *Planta* **2017**, *245*, 737–747. [CrossRef]
7. Evers, A.-M. Effects of different fertilization practices on the carotene content of carrot. *Agric. Food Sci.* **1989**, *61*, 7–14. [CrossRef]
8. Boskovic-Rakocevic, L.; Pavlovic, R.; Zdravkovic, J.; Zdravkovic, M.; Pavlovic, N.; Djuric, M. Effect of nitrogen fertilization on carrot quality. *Afr. J. Agric. Res.* **2012**, *7*, 2884–2900. [CrossRef]
9. Smoleń, S.; Sady, W. The effect of various nitrogen fertilization and foliar nutrition regimes on the concentrations of sugars, carotenoids and phenolic compounds in carrot (*Daucus carota* L.). *Sci. Hortic.* **2009**, *120*, 315–324. [CrossRef]
10. Gajewski, M.; Węglarz, Z.; Sereda, A.; Bajer, M.; Kuczkowska, A.; Majewski, M. Carotenoid Accumulation by Carrot Storage Roots in Relation to Nitrogen Fertilization Level. *Not. Bot. Horti Agrobot. Cluj-Napoca* **2010**, *38*, 71–75. [CrossRef]
11. Legha, M.R.; Prasad, K.V.; Singh, S.K.; Kaur, C.; Arora, A.; Kumar, S. Induction of carotenoid pigments in callus cultures of *Calendula officinalis* L. in response to nitrogen and sucrose levels. *Vitr. Cell. Dev. Biol.-Plant* **2012**, *48*, 99–106. [CrossRef]
12. Saad, K.R.; Kumar, G.; Giridhar, P.; Shetty, N.P. Differential expression of anthocyanin biosynthesis genes in *Daucus carota* callus culture in response to ammonium and potassium nitrate ratio in the culture medium. *3 Biotech* **2018**, *8*, 431. [CrossRef]
13. Baranski, R. Genetic Transformation of Carrot (Daucus carota) and Other *Apiaceae* Species. *Transgenic Plant J.* **2008**, *2*, 18–38.
14. Baranski, R.; Lukasiewicz, A. Genetic Engineering of carrot. In *The Carrot Genome*; Simon, P., Iorizzo, M., Grzebelus, R., Baranski, R., Eds.; Springer International Publishing: Cham, Switzerland, 2019; pp. 149–186. [CrossRef]

15. Iorizzo, M.; Ellison, S.; Senalik, D.; Zeng, P.; Satapoomin, P.; Huang, J.; Bowman, M.; Iovene, M.; Sanseverino, W.; Cavagnaro, P.; et al. A high-quality carrot genome assembly provides new insights into carotenoid accumulation and asterid genome evolution. *Nat. Genet.* **2016**, *48*, 657–666. [CrossRef]
16. Simon, P.W. Classical and Molecular Carrot Breeding. In *The Carrot Genome*; Simon, P., Iorizzo, M., Grzebelus, R., Baranski, R., Eds.; Springer International Publishing: Cham, Switzerland, 2019; pp. 137–147. [CrossRef]
17. Ikeuchi, M.; Sugimoto, K.; Iwase, A. Plant Callus: Mechanisms of Induction and Repression. *Plant Cell* **2013**, *25*, 3159–3173. [CrossRef]
18. Klimek-Chodacka, M.; Oleszkiewicz, T.; Lowder, L.G.; Qi, Y.; Baranski, R. Efficient CRISPR/Cas9-based genome editing in carrot cells. *Plant Cell Rep.* **2018**, *37*, 575–586. [CrossRef] [PubMed]
19. Oleszkiewicz, T.; Klimek-Chodacka, M.; Kruczek, M.; Godel-Jędrychowska, K.; Sala, K.; Milewska-Hendel, A.; Zubko, M.; Kurczyńska, E.; Qi, Y.; Baranski, R. Inhibition of Carotenoid Biosynthesis by CRISPR/Cas9 Triggers Cell Wall Remodelling in Carrot. *Int. J. Mol. Sci.* **2021**, *22*, 6516. [CrossRef] [PubMed]
20. Baranska, M.; Baranski, R.; Schulz, H.; Nothnagel, T. Tissue-specific accumulation of carotenoids in carrot roots. *Planta* **2006**, *224*, 1028–1037. [CrossRef] [PubMed]
21. Baranski, R.; Klocke, E.; Schumann, G. Green fluorescent protein as an efficient selection marker for Agrobacterium rhizogenes mediated carrot transformation. *Plant Cell Rep.* **2006**, *25*, 190–197. [CrossRef]
22. Shimizu, K.; Kikuchi, T.; Sugano, N.; Nishi, A. Carotenoid and Steroid Syntheses by Carrot Cells in Suspension Culture. *Physiol. Plant.* **1979**, *46*, 127–132. [CrossRef]
23. Oleszkiewicz, T.; Klimek-Chodacka, M.; Milewska-Hendel, A.; Zubko, M.; Stróż, D.; Kurczyńska, E.; Boba, A.; Szopa, J.; Barański, R. Unique chromoplast organisation and carotenoid gene expression in carotenoid-rich carrot callus. *Planta* **2018**, *248*, 1455–1471. [CrossRef]
24. Xu, Z.-S.; Feng, K.; Xiong, A.-S. CRISPR/Cas9-Mediated Multiply Targeted Mutagenesis in Orange and Purple Carrot Plants. *Mol. Biotechnol.* **2019**, *61*, 191–199. [CrossRef] [PubMed]
25. Rygula, A.; Oleszkiewicz, T.; Grzebelus, E.; Pacia, M.Z.; Baranska, M.; Baranski, R. Raman, AFM and SNOM high resolution imaging of carotene crystals in a model carrot cell system. *Spectrochim. Acta A* **2018**, *197*, 47–55. [CrossRef] [PubMed]
26. Dudek, M.; Machalska, E.; Oleszkiewicz, T.; Grzebelus, E.; Baranski, R.; Szcześniak, P.; Mlynarski, J.; Zajac, G.; Kaczor, A.; Baranska, M. Chiral amplification in nature: Studying cell-extracted chiral carotenoid microcrystals via the resonance Raman optical activity of model systems. *Angew. Chem. Int. Ed.* **2019**, *58*, 8383–8388. [CrossRef]
27. George, E.F.; Hall, M.A.; Klerk, G.J.D. The Components of Plant Tissue Culture Media I: Macro- and Micro-Nutrients. In *Plant Propagation by Tissue Culture*; George, E.F., Hall, M.A., Klerk, G.J.D., Eds.; Springer International Publishing: Dordrecht, The Netherlands, 2008; pp. 65–113. [CrossRef]
28. Gamborg, O.L.; Miller, R.A.; Ojima, K. Nutrient requirements of suspension cultures of soybean root cells. *Exp. Cell. Res.* **1968**, *50*, 151–158. [CrossRef]
29. Murashige, T.; Skoog, F. A Revised Medium for Rapid Growth and Bio Assays with Tobacco Tissue Cultures. *Phys. Plant.* **1962**, *15*, 474–497. [CrossRef]
30. Hardegger, M.; Sturm, A. Transformation and regeneration of carrot (*Daucus carota* L.). *Mol. Breed.* **1998**, *4*, 119–127. [CrossRef]
31. Grzebelus, E.; Szklarczyk, M.; Baranski, R. An improved protocol for plant regeneration from leaf- and hypocotyl-derived protoplasts of carrot. *Plant Cell Tiss. Organ Cult.* **2012**, *109*, 101–109. [CrossRef]
32. Kiełkowska, A.; Grzebelus, E.; Lis-Krzyścin, A.; Maćkowska, K. Application of the salt stress to the protoplasts cultures of the carrot (*Daucus carota* L.) and the evaluation of the response of regenerants to soil salinity. *Plant Cell Tiss. Organ Cult.* **2019**, *137*, 379–395. [CrossRef]
33. Sugano, N.; Miya, S.; Nishi, A. Carotenoid synthesis in a suspension culture of carrot cells. *Plant Cell Physiol.* **1971**, *12*, 525–531. [CrossRef]
34. Nishi, A.; Kurosaki, F. *Daucus carota* L. (Carrot): In Vitro Production of Carotenoids and Phytoalexins. In *Medicinal and Aromatic Plants V. Biotechnology in Agriculture and Forestry*; Bajaj, Y.P.S., Ed.; Springer: Berlin/Heidelberg, Germany, 1993; pp. 178–191. [CrossRef]
35. Schaub, P.; Rodriguez-Franco, M.; Cazzonelli, C.I.; Álvarez, D.; Wüst, F.; Welsch, R. Establishment of an Arabidopsis callus system to study the interrelations of biosynthesis, degradation and accumulation of carotenoids. *PLoS ONE* **2018**, *13*, e0192158. [CrossRef]
36. Benítez-García, I.; Vanegas-Espinoza, P.E.; Meléndez-Martínez, A.J.; Heredia, F.J.; Paredes-López, O.; Del Villar-Martínez, A.A. Callus culture development of two varieties of Tagetes erecta and carotenoid production. *Electron. J. Biotechnol.* **2014**, *17*, 107–113. [CrossRef]
37. Singh, D.P.; Beloy, J.; McInerney, J.K.; Day, L. Impact of boron, calcium and genetic factors on vitamin C, carotenoids, phenolic acids, anthocyanins and antioxidant capacity of carrots (*Daucus carota*). *Food Chem.* **2012**, *132*, 1161–1170. [CrossRef]
38. Yoshida, A.; Okamura, S.; Sugano, N.; Nishi, A. Effect of Phosphate Concentration on Growth and Carotenoid Synthesis of Carrot Cells in Suspension Culture. *Environ. Control. Biol.* **1975**, *13*, 47–53. [CrossRef]
39. Yun, J.W.; Kim, J.H.; Yoo, Y.J. Optimizations of carotenoid biosynthesis by controlling sucrose concentration. *Biotechnol. Lett.* **1990**, *12*, 905–910. [CrossRef]

40. Hanchincal, V.M.; Survase, S.A.; Sawant, S.K.; Annapure, U.S. Response surface methodology in media optimization for production of β-carotene from *Daucus carota*. *Plant Cell Tiss. Organ Cult.* **2008**, *93*, 123–132. [CrossRef]
41. Irshad, M.; Debnath, B.; Mitra, S.; Arafat, Y.; Li, M.; Sun, Y.; Qiu, D. Accumulation of anthocyanin in callus cultures of red-pod okra [*Abelmoschus esculentus* (L.) Hongjiao] in response to light and nitrogen levels. *Plant Cell Tiss. Organ Cult.* **2018**, *134*, 29–39. [CrossRef]
42. Zheng, H.-Z.; Wei, H.; Guo, S.-H.; Yang, X.; Feng, M.-X.; Jin, X.-Q.; Fang, Y.-L.; Zhang, Z.-W.; Xu, T.-F.; Meng, J.-F. Nitrogen and phosphorus co-starvation inhibits anthocyanin synthesis in the callus of grape berry skin. *Plant Cell Tiss. Organ Cult.* **2020**, *142*, 313–325. [CrossRef]
43. Narayan, M.S.; Venkataraman, L.V. Effect of Sugar and Nitrogen on the Production of Anthocyanin in Cultured Carrot (*Daucus carota*) cells. *J. Food Sci.* **2002**, *67*, 84–86. [CrossRef]
44. Li, S.-X.; Wang, Z.-H.; Stewart, B.A. Chapter Five—Responses of Crop Plants to Ammonium and Nitrate, N. *Adv. Agron.* **2013**, *118*, 205–397. [CrossRef]
45. Kino-oka, M.; Taya, M.; Tone, S. Evaluation of inhibitory effect of ammonium ion on cultures of plant hairy roots. *J. Chem. Eng. Jpn* **1993**, *26*, 578–580. [CrossRef]
46. Okamura, S.; Sueki, K.; Nishi, A. Physiological Changes of Carrot Cells in Suspension Culture during Growth and Senescence. *Physiol. Plant.* **1975**, *33*, 251–255. [CrossRef]
47. Ram, M.; Prasad, K.V.; Kaur, C.; Singh, S.K.; Arora, A.; Kumar, S. Induction of anthocyanin pigments in callus cultures of *Rosa hybrida* L. in response to sucrose and ammonical nitrogen levels. *Plant Cell Tiss. Organ Cult* **2011**, *104*, 171–179. [CrossRef]
48. Kamada, H.; Harada, H. Studies on the organogenesis in carrot tissue cultures II. Effects of amino acids and inorganic nitrogenous compounds on somatic embryogenesis. *Z. Pflanzenphysiol.* **1978**, *91*, 453–463. [CrossRef]
49. Yau, Y.-Y.; Wang, K.J. Increased regeneration ability of transgenic callus of carrot (*Daucus carota* L.) on B5-based regeneration medium. *J. Appl. Hort.* **2012**, *14*, 3–6. [CrossRef]

plants MDPI

Article

Genetic Stability, Phenolic, Flavonoid, Ferulic Acid Contents, and Antioxidant Activity of Micropropagated *Lycium schweinfurthii* Plants

Diaa Mamdouh [1,2,*], Hany A. M. Mahgoub [2], Ahmed M. M. Gabr [3], Emad A. Ewais [2] and Iryna Smetanska [1,*]

1 Department of Plant Food Processing, Agricultural Faculty, University of Applied Sciences Weihenstephan-an-Triesdorf, Markgrafenstr 16, 91746 Weidenbach, Germany
2 Botany & Microbiology Department, Faculty of Science, Al-Azhar University, Nasr City, Cairo 11884, Egypt; h.mahgoub@azhar.edu.eg (H.A.M.M.); ewais_e@yahoo.com (E.A.E.)
3 Department of Plant Biotechnology, Genetic Engineering and Biotechnology Research Division, National Research Centre (NRC), Cairo 12622, Egypt; ahmedgabr01@gmail.com
* Correspondence: diaa.mamdouh@azhar.edu.eg (D.M.); iryna.smetanska@hswt.de (I.S.)

Abstract: *Lycium schweinfurthii* is a Mediterranean wild shrub rich in plant secondary metabolites. In vitro propagation of this plant may support the production of valuable dietary supplements for humanity, introduction of it to the world market, and opportunities for further studies. The presented study aimed to introduce an efficient and reproducible protocol for in vitro micropropagation of *L. schweinfurthii* and assess the genetic stability of micropropagated plants (MiPs) as well as to estimate phenolic, flavonoid, ferulic acid contents, and the antioxidant activity in leaves of micropropagated plants. Two DNA-based techniques, random amplified polymorphic DNA (RAPD) and inter-simple sequence repeats (ISSR), and one biochemical technique, sodium dodecyl sulfate-polyacrylamide gel electrophoresis (SDS-PAGE), were used to assess the genetic stability in MiPs. Spectrophotometric analysis was performed to estimate total phenolic and flavonoid contents and antioxidant activity of MiPs leaves, while ferulic acid content was estimated using high-performance thin-layer chromatography (HPTLC). Sufficient shoot proliferation was achieved at MS (Murashige and Skoog) medium supplemented with 0.4 mg L^{-1} kinetin and rooted successfully on half-strength MS medium fortified with 0.4 mg L^{-1} Indole-3-butyric acid (IBA). The Jaccard's similarity coefficients detected in MiPs reached 52%, 55%, and 82% in the RAPD, ISSR, and SDS-PAGE analyses, respectively. In the dried leaves of MiPs, the phenolic, flavonoid, and ferulic acid contents of 11.53 mg gallic acid equivalent, 12.99 mg catechin equivalent, and 45.52 mg were estimated per gram, respectively. However, an IC_{50} of 0.43, and 1.99 mg mL^{-1} of MiP dried leaves' methanolic extract was required to scavenge half of the DPPH, and ABTS free radicals, respectively. The study presented a successful protocol for in vitro propagation of a valued promising plant source of phenolic compounds.

Keywords: *Lycium schweinfurthii*; micropropagation; genetic stability; ISSR-PCR; RAPD-PCR; SDS-PAGE; HPTLC; DPPH; ABTS

Citation: Mamdouh, D.; Mahgoub, H.A.M.; Gabr, A.M.M.; Ewais, E.A.; Smetanska, I. Genetic Stability, Phenolic, Flavonoid, Ferulic Acid Contents, and Antioxidant Activity of Micropropagated *Lycium schweinfurthii* Plants. *Plants* **2021**, *10*, 2089. https://doi.org/10.3390/plants10102089

Academic Editors: Laura Pistelli and Kalina Danova

Received: 7 September 2021
Accepted: 28 September 2021
Published: 1 October 2021

1. Introduction

One member of the Solanaceae (the nightshade) family is the genus Lycium, comprising more than 70 species and which has a disjunctive distribution in temperate to subtropical regions in South America, North America, Africa, Eurasia, and Australia [1]. Within buckthorns (Lycium), *Lycium schweinfurthii* is grouped according to phylogenetic studies in a clade with other Old World species of the genus. Within this clade, this species is closely related to *L. acutifolium*, *L. eenii*, *L. shawii*, *L. bosciifolium*, *L. hirsutum*, and *L. villosum*. The species is sometimes put to *L. intricatum* [2]. *L. schweinfurthii* grows in temperate climates and is well spread throughout the southern Mediterranean region as well as in Egypt, Algeria, Tunisia, and Libya [3]. *L. schweinfurthii* is distributed in Egypt in the great south-western desert, northern coastal region [4], and islands of Lake Burullus [3]. The plant is a 2–3 m high, rigid, upright shrub with a spiny stem. Its leaves are succulent and hairless that are 12–20 mm long and 2–4 mm wide and arranged in alternate patterns (one leaf per node) while its flowers are hermaphrodite. The fruit is a black, spherical, sometimes egg-shaped berry that measures 4–5 mm in diameter [5]. *L. schweinfurthii* suffers from different types of threats that affect its distribution, whether natural or caused by human activities, i.e., soil fragmentation, cutting, grazing, and firing [3].

It is difficult in many seasons to obtain seeds or crops from wild plants, especially with their small number and wide geographical distribution, as in *L. schweinfurthii*. Hence, it is imperative to micropropagate plants in vitro to maintain the explant source at all times of the year. For decades, the micropropagation of plants was the only technique that maintained and promoted the economic value of many agricultural species [6]. Furthermore, it is an efficient technique for in vitro multiplication of endangered species, e.g., *Magnolia sirindhorniae* [7], as well as for producing secondary metabolites, e.g., *Eryngium alpinum* L. [8]. Although no reports were found on the micropropagation of *L. schweinfurthii*, it is well studied in other species of the Lycium genus. Multiple shoots and adventitious buds of *L. ruthenicum* were developed in vitro not only from stems but also from leaf explants [9]. Moreover, the best shoot proliferation of *L. depressum* was achieved at a low concentration of BA (6-benzyl adenine) and rooted in full-strength MS medium (Murashige and Skoog medium) supplemented with IBA (indole-3-butyric acid) with a high survival rate [10]. Micropropagation protocols were also developed in *L. barbarum* [11] and *L. chinense* [12].

To maintain the effectiveness of in vitro propagation, genetic stability must be ensured, especially with successive generations. Diverse techniques are used to determine the genetic stability of regenerated plants in terms of plant genomes or transcribed proteins. One of these is the random amplified polymorphic DNA (RAPD) PCR technique, which is a rapid, inexpensive, and simple method for detecting genetic differences as it does not require any previous information about the plant genome [13]. RAPD-PCR was used to determine the genetic stability in micropropagated plants of *Prunus salicina* [14], *Echinacea purpurea* [15], *Dendrobium fimbriatum* [16], and *Rhynchostylis retusa* [17]. A more specific technique than RAPD is the inter-simple sequence repeats (ISSR) PCR technique. It is an efficient, quick, and reproducible technique in which the targets are the DNA fragments located between adjacent microsatellite regions, while the RAPD-PCR targets are random [18]. Wójcik et al. [19] used ISSR primers to observe the genetic stability of regenerated plants of *Ribes grossularia* L. Both techniques are used together to obtain more realistic and accurate results [20–22].

Otherwise, the differences in the protein profile of the regenerated plants also reflect the extent of genetic stability at the level of gene expression. Sodium dodecyl sulfate-polyacrylamide gel electrophoresis (SDS-PAGE) is a technique used to show the differences in the transcribed polypeptides in the micropropagated plants concerning the mother plant [23,24]. The SDS-PAGE technique was assessed to check the genetic stability of in vitro micropropagated plants of *Pilosocereus robinii* [24], *Musa* spp. [25], and *Phoenix dactylifera* L. [26]. At both levels, the DNA genome and the transcribed proteins are essential for recognizing the genetic stability in the regenerated plants of *L. schweinfurthii*.

The functional effect of certain plant species and their use in folk medicine depend mainly on their active secondary metabolites [27,28]. Plants of *L. schweinfurthii* have been reported to contain a high level of phenolic compounds, particularly flavonoids [29]. These secondary metabolites play a major role in adapting the plant to the environment and maintaining its survival [30]. Flavonoids are naturally produced phenolic compounds in plants and play an important role in the protection against unfavorable environmental conditions such as drought [31], high concentrations of aluminum in soil [32], UV-irradiation [33], and defense plants against herbivores, bacteria, and fungi [34]. Phenolic compounds have a role in modern human therapy, e.g., controlling hyperglycemia associated with type 2 diabetes at early stages when included in the human diet [35]. Moreover, flavonoids are reported to protect humans against numerous diseases due to the fact of their strong anti-oxidative [36], anti-inflammatory [37], anticarcinogenic [38], antiviral [39], and antibacterial [40] activities as well as a direct cytoprotective effect on several human systems (i.e., coronary and vascular systems) and organs (i.e., liver and pancreas) [41,42]. These features put them among the most attractive natural substances available for enhancing the options of the previously mentioned therapy [43]. The leaves of *L. schweinfurthii* contain large quantities of flavonoids compared to roots, stems, and flowers [44]. The main phenolics found in leaves are quercetin, kaempferol, gallic acid, ferulic acid, and apigenin [29]. Six glucosides have been isolated from *L. schweinfurthii*. Four of them showed a potent inhibitory activity that could decrease postprandial hyperglycemia in diabetic patients [45].

Although many plants contain high-value phenolic compounds, it is difficult to cultivate at a large-scale due to the specific ecological conditions. Corresponding plant in vitro cultures, particularly plant cell cultures, provide an attractive alternative source of phenolics that can overcome the limitations of extracting useful metabolites from limited natural resources [46]. Obtaining phenolic compounds from plant's in vitro cultures is one of the more interesting research areas in recent decades due to the fact of their benefits. Phenolic content can be elevated in culture medium such as in *Zingiber officinale* Rosc. [47], *Sequoia sempervirens* [48], *Rosa damascene* Mill [49], and grape [50].

It is worth searching for alternative plant sources to meet the nutritional needs of humans and to protect them from diseases resulting from malnutrition and a lack of functional substances in the future. Thus, the present study is the first attempt to optimize a protocol for direct in vitro plant regeneration in *L. schweinfurthii* as well as to evaluate its phenolic, flavonoid, ferulic acid contents, and antioxidant activity of in vitro leaves' extract.

2. Materials and Methods

2.1. Plant Material and Culture Conditions

Fruits of *L. schweinfurthii* were collected during March 2016 from Jazirat Al-Kawm Al-Akhdar (the green islet) located in Burullus Lake (northern Nile Delta), Egypt. The fruits were air-dried for approximately 120 h, and then their envelopes were removed to obtain their seeds. The plant seeds were washed with 70% ethanol for 30 s, and then they were surface sterilized by soaking in 30% commercial Clorox for 10 min. Seeds were washed with sterilized distilled water 4 times to remove the remaining bleach.

After the sterilization process, seeds were cultured in 300 mL jars containing 30 mL basal MS medium, including vitamins (Caisson Labs, Smithfield, UT, USA), with 3% sucrose and solidified using 7% agar (ROTH Company, Carlsruhe, Germany) and incubated at 23 ± 2 °C under a 16 h photoperiod of 2500 lux by cool fluorescent lamps.

2.2. In Vitro Micropropagation

For vegetative propagation, nodal segments were cut and cultivated on full-strength MS media including vitamins supplemented with BA (0.4, 0.8, 1.6, or 3.2 mg L^{-1}), kinetin (0.4, 0.8, 1.6, or 3.2 mg L^{-1}), BA + Kin (0.2 + 0.2, 0.4 + 0.4, 0.8 + 0.8, or 1.6 + 1.6 mg L^{-1}), or BA + Kin + NAA (0.2 + 0.2 + 0.2 or 0.4 + 0.4 + 0.4 mg L^{-1}) and on basal MS medium as a control. Seven nodal explants were used for shoot formation in each treatment. Regenerated shootlets were then transferred to basal full-strength MS, half-strength MS, half-strength MS medium fortified with NAA (0.4, 0.8, or 1.6 mg L^{-1}) or IBA (0.4, 0.8, or 1.6 mg L^{-1}). To determine the rooting capacity and the most suitable rooting medium, eight shootlets were used in each treatment.

2.3. DNA Extraction and PCR Amplification Conditions

Total DNA was extracted from leaves of two in vitro mother plants and their micropropagated plantlets for three generations using the E.Z.N.A. kit (VWR, Darmstadt, Germany). Twelve primers (i.e., 7 RAPD and 5 ISSR) out of a total of twenty primers (Thermo Fisher, Frankfurt, Germany) were selected to amplify DNA fragments. The protocol for RAPD and ISSR analysis was adapted from Martins et al. [51] and Williams et al. [52]. PCR was performed in a volume of 20 µL using Invitrogen™ Platinum™ master mix (Thermo

Fisher, Frankfurt, Germany). The amplification reaction consisted of an initial denaturation step at 94 °C for 5 min, followed by 43 cycles of 1 min at 92 °C, 1 min at a specific annealing temperature (Table 1), and 2 min at 72 °C; there was one last extension step of 7 min at 72 °C. Amplifications were performed in a Bio-Rad T100™ thermal cycler (Bio-Rad Laboratories, Hercules, CA, USA) for both RAPD and ISSR. DNA amplification fragments were separated with 1.5% agarose gel using 1x TBE buffer and stained with Red-Safe™ nucleic acid staining solution. Gels were then analyzed with CAMAG® TLC Visualizer 2 (CAMAG, Muttenz, Switzerland).

Table 1. Sequences and annealing temperatures of primers used for the RAPD and ISSR analysis of *L. schweinfurthii*.

Primer Name		Sequence	Annealing Temperature (°C)
RAPD primers	OPA-10	5′-GTGATCGCAG-3′	40.5
	OPAJ-01	5′-ACGGGTCAGA-3′	43
	OPAK-06	5′-TCACGTCCCT-3′	42
	OPAK-20	5′-TGATGGCGTC-3′	41
	OPAQ-20	5′-GTGAACGCTC-3′	40.5
	OPB-18	5′-CCACAGCAGT-3′	42
	OPR-09	5′-TGAGCACGAG-3′	42
ISSR primers	HB11	5′-GTGTGTGTGTGTCC-3′	54
	HB12	5′-CACCACCACGC-3′	50.9
	HB13	5′- GAGGAGGAGGC-3′	48
	HB14	5′-CTCCTCCTCGC-3′	48
	HB15	5′-GTGGTGGTGGC-3′	50.9

2.4. Protein Extraction and SDS-PAGE

Total protein was extracted from the healthy leaves of two in vitro mother plants and their micropropagated plantlets for three generations. Ten milligrams of ground, fine powder were homogenized thoroughly with a 400 µL extraction buffer (0.6 g Tris base, 0.2 g sodium dodecyl sulfate (SDS), 30 g of urea, and 1 mL β-mercaptoethanol in 100 mL double-distilled water) using vortex. The mixture was centrifuged at 13,000 rpm for 10 min at room temperature after keeping overnight at 4 °C. Twenty microliters of the extracted protein samples were boiled in a water bath for 3–5 min before loading them on the gel. SDS-PAGE was performed according to Laemmli [53] using 12.5% resolving gel, 4% stacking gel, and bromophenol blue as a tracking dye. After carrying out the electrophoresis at 150 volts and 25 milliamperes, the gel was de-stained in a methanol:glacial acetic acid:water (4:1:5) mixture. Then, it was kept overnight in Coomassie Brilliant Blue buffer for staining. The gel was photographed, and the molecular weights of the polypeptide bands were estimated against protein molecular weight marker.

2.5. Secondary Metabolites

2.5.1. Sample Preparation and Extraction

Leaves of micropropagated plants were randomly collected, freeze-dried, and ground. One gram was collected and Soxhlet extracted with 200 mL of 80% aqueous methanol for 24 h. The extract was concentrated with a rotary evaporator to a concentration of 50 mg mL^{-1} which was then subjected to estimate the phenolic and flavonoid contents as well as the antioxidant activity. More diluted leaves' extract of 10 mg mL^{-1} was used to quantify the ferulic acid content through HPTLC (high-performance thin layer chromatography) analysis.

2.5.2. Total Phenolic Assay

The total phenolic content of the leaves was determined using the Folin–Ciocalteu assay as described by Marinova et al. [54] with some modifications. An aliquot (200 µL) of extracts or gallic acid (Sigma–Aldrich, St. Louis, MO, USA) standard solution (10, 20, 30, 40, 50, and 100 mg L^{-1}) was added to a 5 mL Eppendorf tube containing 1.8 mL distilled deionized water. Two hundred microliters of Folin-Ciocalteu's reagent (Merck, Schnelldorf, Germany) were added to the mixture and shaken. After 5 min, 2 mL of 7% sodium carbonate (VWR chemicals, Darmstadt, Germany) solution was added and mixed thoroughly. The mixture was diluted to 5 mL with distilled water and incubated for 90 min in the dark at room temperature. The absorbance against the reagent blank was determined at 750 nm with an Analytic Jena Specord® 250 Plus UV-Vis spectrophotometer. Total phenolic content is expressed as mg GAE g^{-1} DW (mg gallic acid equivalents/g dry weight) and calculated as follows: T = CV/M, where T is the total phenolic content, C is the concentration of gallic acid estimated in mg mL^{-1}, V is the volume of extract solution in mL, and M is the weight of extract in g.

2.5.3. Total Flavonoid Assay

Total flavonoid content was measured using the aluminum chloride assay as described by Marinova et al. [54] with some modifications. An aliquot (500 µL) of extracts or catechin standard (Sigma–Aldrich, St. Louis, MO, USA) solution (10, 20, 30, 40, 50, and 100 mg L^{-1}) was added to a 5 mL Eppendorf tube, containing 2 mL distilled water. To the diluted sample, 150 µL of 5% sodium nitrite (AppliChem, Darmstadt, Germany) was added. After 5 min, 150 µL of 10% aluminum chloride (Carl-Roth, Carlsruhe, Germany) was added. At the sixth min, 1 mL of 1 M sodium hydroxide was added, and the total volume was diluted to 5 mL using distilled water. The absorbance was measured against reagent blank at 510 nm, and total flavonoids were expressed as mg CE g^{-1} DW (mg catechin equivalent/g dry weight) and calculated by the equation: T = CV/M, where T is the total flavonoid content, C is the concentration of catechin estimated in mg mL^{-1}, V is the volume of extract solution in ml, and M is the weight of extract in g.

2.5.4. HPTLC Conditions

The high-performance thin-layer chromatography (HPTLC) system (Camag, Muttenz, Switzerland) consisted of a Limomat 5 connected to compressed air, an Automatic Developing Chamber 2 (ADC 2), and a TLC Visualizer 2 supported with visionCATS software. An analytical grade of ferulic acid (Merck, Germany) was used to prepare 400 µg ml^{-1} in methanol as a calibration standard against dry leaves' extracts of micropropagated plants. TLC silica gel 60 F_{254} aluminum plates (10 × 20 cm, Merck, Darmstadt, Germany) were used for the TLC analysis. Standard and samples were applied to plates as 8 mm bands, 8 mm from the bottom edge of the layer, using Linomat 5. A ferulic acid standard solution of 400 µg ml^{-1} of a volume of 2–9 µL was applied against 2, 4, 6, 8, 10, 12, and 14 µL of dry leaves' extract. A mixture of ethyl acetate/methanol/water (100:13.5:10, *v*/*v*/*v*) was used as the mobile phase. Plates were developed at room temperature and 60% humidity in an ADC2 automated development chamber. The migration distance of the mobile phase was 70 mm with a development time of 9 min. After development, the chromatogram was visualized and photographed by Visualizer 2 at 254 and 366 nm. The ferulic acid content in the samples was expressed as mg g^{-1} DW.

2.6. *Antioxidant Capacity*

The antioxidant capacity of the micropropagated leaves' extract was measured using the DPPH (diphenyl-1-picryl-hydrazyl) assay according to Olalere et al. [55] and the ABTS (2,2′-azino-bis(3-ethylbenzothiazoline-6-sulphonic acid)) assay according to Gabr et al. [56].

2.6.1. DPPH Free Radical Scavenging

DPPH is highly sensitive in detecting small differences in antioxidant activities. It is a stable free radical that can accept a hydrogen radical or an electron to convert to a stable molecule. The stock solution of DPPH reagent (1 mM) was prepared and stored at –20 °C until use. The working solution (0.06 mM) was prepared to obtain an absorbance value of 0.8 ± 0.04 at 515 nm. Ten different extract concentrations of micropropagated leaves (between 0.25 and 0.7 mg mL^{-1}) were prepared. The absorbance at 515 nm (A_1)

was measured for a mixture of 0.5 mL of each extract concentration with 2.5 mL DPPH working solution after incubation in the dark at room temperature for 30 min. Ethanol was used instead of extract to obtain the absorbance of the control reaction (A_0). The DPPH radical scavenging activity percentage was calculated as follows: $((A_0 - A_1)/A_0) \times 100$. The inhibition percentage was plotted against the different concentrations of the leaves' extracts to generate a straight-line equation. The extract concentration required to scavenge half of the DPPH radicals (IC_{50}) was then determined.

2.6.2. ABTS Free Radical Scavenging

A 7 mM ABTS solution was reacted with 2.4 mM potassium persulphate solution at a ratio of 1:1 (*v*/*v*). The solution was incubated in the dark at room temperature for 16 h. One milliliter of the prepared $ABTS^+$ solution was diluted with 60 mL methanol resulting in a working solution with an absorbance of 0.60 ± 0.01 at 728 nm. Fourteen different extract concentrations of micropropagated leaves (between 1.0 and 5.5 mg mL^{-1}) were prepared. The absorbance at 734 nm (A_1) was measured for a mixture of 40 µL of each extract concentration with 1.96 mL blue-green $ABTS^+$ working solution after incubation in the dark at 37 °C for 10 min. The control reaction (A_0), which contains all reagents except the test compound, was run identically. The $ABTS^+$ radical scavenging activity percentage was calculated as follows: $((A_0 - A_1)/A_0) \times 100$. The inhibition percentage was plotted against the different concentrations of the dry leaves' extracts to generate a straight-line equation. The concentration of extract required to scavenge half of the $ABTS^+$ radicals (IC_{50}) was then determined.

2.7. *Recording Data and Statistical Analysis*

The number of plantlets, leaves, distinct nodes, and shootlet length were estimated and recorded after five weeks of cultivation. Recorded data were subjected to statistical analysis of variance (ANOVA) using SigmaPlot v.12.5. The Shapiro–Wilk normality test failed for all data and also for the transformations in the number of plantlets, nodes, roots, and root length. Then, the power of the performed test decreased from 0.50 to 0.001. The normality test passed in shootlet length and passed in the number of leaves after transformation into the square root. The Holm–Sidak method was applied for pairwise comparisons.

RAPD, ISSR, and SDS-PAGE data were scored for presence (1) and absence (0). Three matrices were generated, one for each analysis type. The genetic similarities were calculated according to Jaccard's index. A dendrogram showing the genetic stability between the three generations' individuals and the mother plant was constructed using UPGMA (unweighted pair group method with arithmetic average) through CAP 1.2 software [57].

3. Results and Discussion

3.1. *In Vitro Propagation*

During the study of in vitro seed germination of *L. schweinfurthii*, the percentage of microbial contamination based on the method of sterilization used in culture media was 16.67%, and the maximum percentage of germination in non-contaminated cultures was 30% (Figure 1). Although due to the plant's relatively low germination percentage, it was noticed that the germinated plants had plenty of leaves, convergent nodes, and elongated stems. The low growth rate reflected the low spread of the plant across large areas in the wildlife, which means that the in vitro multiplication of the plant is of great significance.

In an attempt for intensive plant micropropagation, nodal segments of sterilized germinated seedlings were cut and transferred into full-strength MS medium fortified with different concentrations of BA, Kin, and NAA as explained in Table 2. Strong variability was obtained in the number of leaves and shoot length of regenerated plantlets after 5 weeks of culture.

Figure 1. In vitro plant micropropagation protocol of *L. schweinfurthii*: (**a**) aseptic seedling; (**b**) shoot formation after five weeks of culture on MS medium supplemented with 0.4 mg L^{-1} Kinetin; (**c**) roots formed on MS medium fortified with 0.4 mg L^{-1} IBA (indole-3-butyric acid).

Table 2. Effect of different concentrations of BA, Kin, and NAA on micropropagation of *L. schweinfurthii* from nodal cuttings.

Treatment (mg L^{-1})			Number of Plantlets	Number of Distinct Nodes	Number of Leaves	Shootlet Length (cm)
BA	Kin	NAA				
-	-	-	1.00 ± 0.00^{abc}	1.00 ± 0.00^{abc}	4.14 ± 0.46^{c}	0.56 ± 0.03^{b}
0.4	-	-	1.43 ± 0.48^{abc}	3.29 ± 0.89^{ab}	20.71 ± 3.66^{ad}	1.24 ± 0.40^{b}
0.8	-	-	1.00 ± 0.22^{abc}	1.71 ± 0.47^{abc}	5.86 ± 2.19^{cd}	0.83 ± 0.17^{b}
1.6	-	-	1.14 ± 0.14^{abc}	1.29 ± 0.18^{abc}	7.14 ± 2.44^{bcd}	0.71 ± 0.19^{b}
3.2	-	-	0.29 ± 0.18^{c}	0.43 ± 0.30^{b}	1.86 ± 1.70^{c}	0.20 ± 0.13^{b}
-	0.4	-	1.86 ± 0.46^{ab}	5.86 ± 0.91^{a}	26.00 ± 4.34^{a}	2.83 ± 0.39^{a}
-	0.8	-	0.86 ± 0.14^{abc}	1.71 ± 0.52^{abc}	8.00 ± 2.17^{bcd}	0.93 ± 0.25^{b}
-	1.6	-	0.14 ± 0.14^{c}	0.14 ± 0.14^{b}	0.43 ± 0.43^{c}	0.07 ± 0.07^{b}
-	3.2	-	0.86 ± 0.14^{abc}	1.00 ± 0.22^{abc}	8.14 ± 1.97^{bcd}	0.64 ± 0.13^{b}
0.2	0.2	-	2.00 ± 0.44^{a}	3.86 ± 1.12^{a}	22.14 ± 4.49^{ab}	1.23 ± 0.20^{b}
0.4	0.4	-	1.29 ± 0.18^{abc}	2.14 ± 0.55^{abc}	12.71 ± 4.20^{a}	1.11 ± 0.24^{b}
0.8	0.8	-	1.43 ± 0.30^{abc}	2.14 ± 0.46^{abc}	12.43 ± 3.61^{a}	1.14 ± 0.31^{b}
1.6	1.6	-	1.00 ± 0.22^{abc}	1.71 ± 0.61^{abc}	8.29 ± 2.73^{bcd}	0.91 ± 0.26^{b}
0.2	0.2	0.2	0.57 ± 0.37^{c}	1.00 ± 0.66^{abc}	6.00 ± 3.93^{cd}	0.51 ± 0.34^{b}
0.4	0.4	0.4	0.43 ± 0.20^{c}	1.29 ± 0.97^{abc}	4.43 ± 2.41^{c}	0.63 ± 0.38^{b}

Pairwise comparison was conducted according to the Holm–Sidak method at $p \leq 0.05$. Seven replicates were used for each treatment; BA, 6-benzyl adenine; Kin, kinetin; NAA, naphthalene acetic acid. The letters a, b, c, and d represent the pairwise comparison and the significance between treatments.

The highest significant results of shootlet length were observed in plantlets produced in MS medium with 0.4 mg L^{-1} kinetin, while 0.4 BA, 0.4 kinetin, and 0.2 BA + 0.2 Kin (in mg L^{-1}) were recorded as highly significant in the number of distinct nodes (Figure 1). Although 0.4 mg L^{-1} kinetin was non-significant in other variables, with most treatments used it was the best in terms of the average number of leaves at approximately 26 leaves per regenerated plant. Moreover, it was second (1.86 plantlets/nodal segment) after 0.2 + 0.2 mg L^{-1} Kin + BA in terms of the number of plantlets regenerated per inoculated cut (2 plantlets/ nodal segment).

In vitro propagation of plants depends mainly on the addition of cytokinins to culture media and, sometimes, in addition to a lower concentration of auxins [58]. Two cytokinins

(BA and Kin) and one auxin (NAA) were used for multiple shoot formations from nodal segments of *L. schweinfurthii*. The lower concentrations of cytokinins (BA or Kin) were the best in all determined variables, such as the number of plantlets, nodes, leaves, and shoot length. In the present study, a reduction in shoot proliferation by increasing benzyladenine or kinetin in the culture medium was noticed. Furthermore, similar results were observed when combinations between both growth regulators were added but with a total concentration the same as the concentration of only one of them. This allows saying that shoot formation in *L. schweinfurthii* may depend more on the concentration of the hormone than its type. These results are different from results obtained in the micropropagation of *Magnolia sirindhorniae*, *Eryngium alpinum*, and *Argania spinosa*. Shoots of *M. sirindhorniae* were optimally induced in a half-strength MS medium supplemented with a combination of BA, NAA, and gibberellic acid (GA_3) with higher concentrations, i.e. 2.0 + 0.1 + 2.0 mgL^{-1}, respectively [7]. A solid MS medium combined with BA, IAA, and GA_3 was successful in shoot proliferation of *E. alpinum* [8]. Moreover, the highest adventitious shoots of the endangered plant, *A. spinosa*, were observed on MS medium containing 1 mg L^{-1} BA and 2 mg L^{-1} GA_3 [59].

For completing in vitro micropropagation of the studied species, shootlets of the plant were transferred firstly to full- and half-strength MS media without growth regulators for root initiation. It was noticed that the number and length of roots that emerged in 1/2 MS were better than in the full-strength MS. Therefore, the experiment was repeated with the same treatments in addition to adding NAA and IBA to half-strength MS for rooting enhancement. It was found that increasing NAA concentration to medium reversely affected rooting production. Otherwise, the addition of 0.4 mg L^{-1} IBA enhanced the number of roots and the root length but with non-considerable significance with other treatments according to pairwise comparison using the Holm–Sidak method (Figure 1). Although the highest mean of the number of roots emerged per plant and long roots obtained in IBA treatments, not all eight shootlets showed a rooting response to the treatment. This led to a high standard error in several treatments and hid the significant differences between the different IBA concentrations used (Figure 2). However, the IBA treatments showed significantly better root formation and enhancement than NAA treatments.

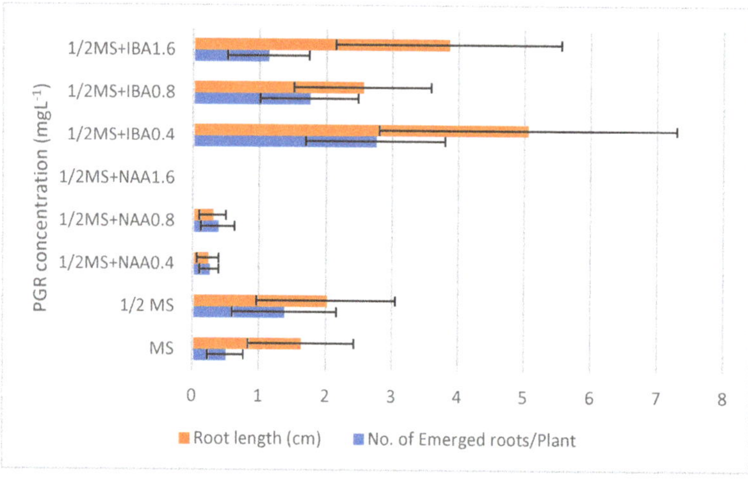

Figure 2. In vitro rooting of *L. schweinfurthii* shoots on MS medium fortified with different auxins. Pairwise comparison showed no significant differences between treatments at $p \leq 0.05$ using the Holm–Sidak method. Eight replicates were used in each treatment. PGR, plant growth regulators; MS, full-strength MS salts; 1/2MS, half MS salts; IBA, indole-3-butyric acid; NAA, naphthalene acetic acid.

In this study, half-strength MS medium with NAA and IBA were used for root stimulation, and IBA was the best for root formation enhancement. The results were consistent with other studies where IBA stimulated sufficient root induction in several species including *Cardiospermum halicacabum* [60], *Dorem ammoniacum* [61], *Achyranthes aspera* [62], and *Prunus armeniaca* L. [63].

3.2. Genetic Stability of Micropropagated Plantlets

For determining the genetic stability in the suggested micropropagation protocol, RAPD, ISSR, and SDS-PAGE analyses were performed to compare between the in vitro mother plant and its micropropagated plantlets, which resulted from using MS medium fortified with 0.4 mg L^{-1} BA for three generations and two individuals from each generation. Among the 20 primers screened (10 RAPD and 10 ISSR), only 12 primers produced clear and detectable amplified DNA fragments and were used in further PCR analysis.

With seven RAPD primers, 29 DNA fragments (a total of 137 scorable bands) were amplified in the mother plant and its three generations plantlets. Jaccard's similarity coefficient, ranging between 0.36 and 0.56, was obtained. The second and third generations showed a similarity of 0.48 and 0.52 to the mother plant, respectively. The highest polymorphism of 100% was observed in fragments amplified with OPA10 and OPAJ01 primers, while the lowest of 50% was in the amplified fragments using OPB18 primer. Only eight monomorphic fragments out of 29 DNA fragments were recorded. Furthermore, 28 DNA fragments (a total of 107 scorable bands) were amplified using five ISSR primers, while a similarity of 0.33-0.70 was recorded. The highest similarities to the mother plant were in the first (0.66) and third (0.55) generations (Figure 3). A higher polymorphism was observed over RAPD, where the lowest was 75% in the HB13 and HB14 primers and the highest was 100% in the HB11 primer. Out of 28 DNA fragments amplified with five ISSR primers, only four fragments were monomorphic.

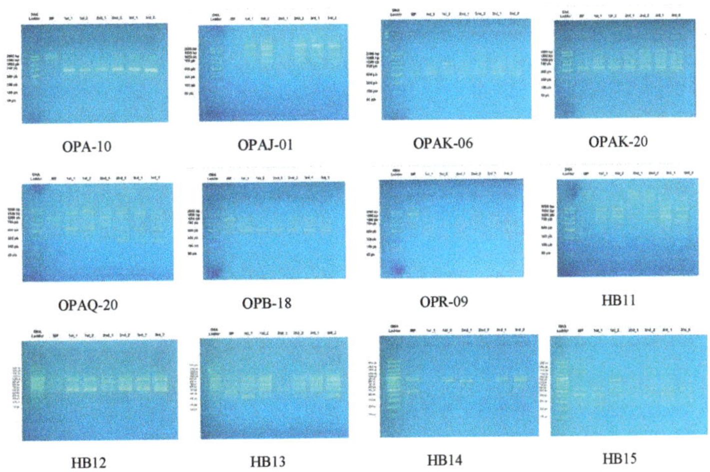

Figure 3. RAPD and ISSR profiles with primers mentioned in Table 1 of three micropropagated generations of *L. schweinfurthii* compared to the mother plant. MP, mother plant; 1st, first-generation plantlets; 2nd, second-generation plantlets; 3rd, third-generation plantlets.

In SDS-PAGE analysis, sixteen polypeptides were separated with a similarity between 0.54 and 0.82. The first and second generations showed high similarity to the mother plant of 0.68 and 0.74, respectively (Figure 4). Half of the separated polypeptides were monomorphic, as they were found in all protein extracts. It was noticed also that there were two unique polypeptides of 82 and 108 KDa that were separated only in a plant in the third generation (3rd_1). The expressed protein showed uniformity between the mother plant and most of the plant individuals studied. On the contrary, only eight polypeptides were separated from *L. schweinfurthii* seed proteins in the study by El-Ghamry et al. [64].

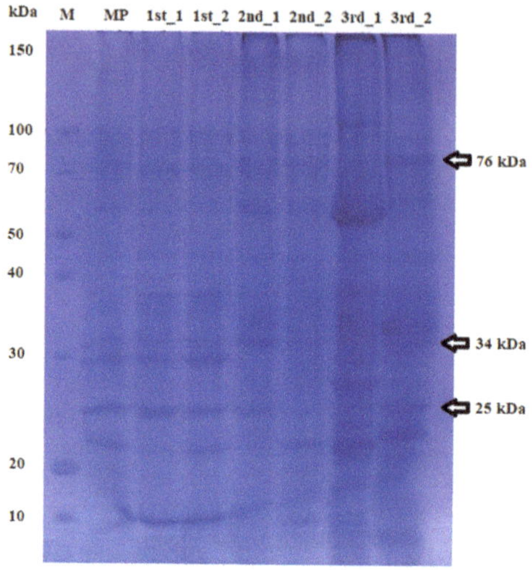

Figure 4. SDS-PAGE analysis of total protein bans extracted from three micropropagated generations of *L. schweinfurthii* compared to the mother plant. M, marker; MP, mother plant; 1st, first-generation plantlets; 2nd, second-generation plantlets; 3rd, third-generation plantlets showing three of the monomorphic polypeptides detected.

Three matrices of RAPD, ISSR, and SDS-PAGE were merged and analyzed to show the clonal fidelity of the DNA and protein levels together. The dendrogram of genetic distances among the in vitro and micropropagated plants based on amplified DNA fragments generated by RAPD and ISSR primers and polypeptides separated in SDS-PAGE is shown in Figure 5. The distances in the dendrogram revealed that the first and third generations of the first plant individuals (1st_1 and 3rd_1) were more similar than the second generation (2nd_1). Furthermore, the second generation of the second individual (2nd_2) was more similar to the mother plant than the first (1st_2) and third (3rd_2) generations. The results showed that the generation that was more similar to the in vitro plants had the higher Jaccard's similarity coefficient which ranged between 0 (completely different) and 1 (identical). The first micropropagated generation showed a higher similarity coefficient to the mother in vitro plants of 0.56–0.58. On the other hand, the second generation showed a similarity coefficient of 0.44–0.61, while the third one showed a similarity coefficient of 0.52–0.56 (Table 3). It was also obtained that the conditions of propagation in this study lowered the tendency of the plants to be genetically stable.

It is necessary after micropropagation to check the genetic uniformity of micropropagated plantlets [65]. Two PCR-based techniques (RAPD and ISSR) and a biochemical marker technique (SDS-PAGE) were used in the present study to test the genetic stability and polypeptide content because of their rapidity, simplicity, and effectiveness as well as the fact that they do not need prior information about the DNA sequence [66]. Moreover, the use of different markers in parallel provides better opportunities for genetic alteration identification between different clones [67]. The molecular markers were not affected by external environmental factors which, consequently, accurately detected the genetic variability among the plant clones [68]. The advantage of using both biochemical and molecular markers is the ability to give an account of the expression stability level of the DNA regarding the variability that occurred in the plant genome. In the present investigation, it was concluded that molecular and biochemical markers are equally important for genetic analysis and for the evaluation of the amount of genetic variability among the different micropropagated plantlets of *L. schweinfurthii*. In addition, Osman et al. [69] determined the genetic relationship between several species of *Zea mays* and *Sorghum* using SDS-PAGE of seed protein as well as RAPD-PCR markers.

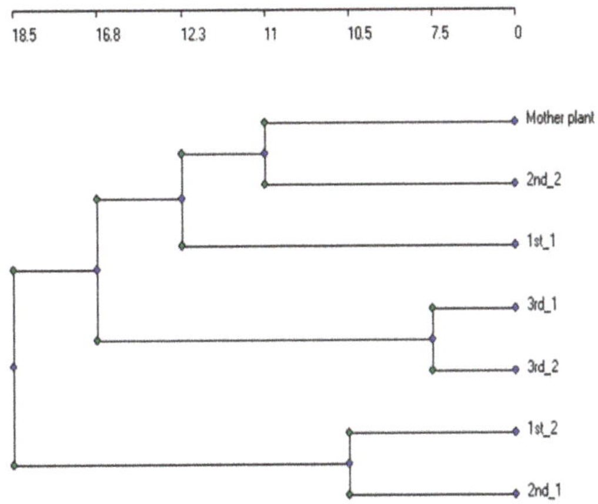

Figure 5. UPGMA dendrogram based on data generated from biochemical and molecular markers, showing the genetic linkage distance among the different micropropagated plantlets in different generations of *L. schweinfurthii*. MP, mother plant; 1st, first-generation plantlets; 2nd, second-generation plantlets; 3rd, third-generation plantlets.

Table 3. Jaccard's similarity coefficient concerning similarities in DNA fragments generated in RAPD and ISSR analyses and protein polypeptides through SDS-PAGE.

	Mother Plant	1st_1	1st_2	2nd_1	2nd_2	3rd_1	3rd_2
Mother Plant							
1st_1	0.5806						
1st_2	0.5593	0.5738					
2nd_1	0.4364	0.4821	0.5714				
2nd_2	0.614	0.6271	0.5	0.54			
3rd_1	0.5625	0.6508	0.5556	0.4912	0.6333		
3rd_2	0.5246	0.5645	0.5424	0.4717	0.569	0.7368	

In the present analysis, SDS-PAGE revealed the high stability of expressed proteins in the micropropagated plantlets compared to the amplified DNA fragments assessed by RAPD- and ISSR-PCR techniques. This indicates that it was supposed to have modifications in plantlet DNA, especially in the non-coding region. This effect may be related to the PGR used in micropropagation, as it was noticed that 6-benzyl adenine affects DNA and causes mutations [70]. In a study by Alizadeh and Singh [71], the similarity coefficient was 1 (in both RAPD and ISSR) in most clones, although there were low coefficients of 0.53 (RAPD) and 0.63 (ISSR) recorded in some clones of *Vitis* spp. micropropagated plantlets. This also raises the idea of the effects of PGRs and the cultivation conditions on the genetic stability of cloned plants.

3.3. Phenolic and Flavonoid Content Estimation

The phenolic and flavonoid contents of the micropropagated plant leaves' extract were estimated spectrophotometrically in terms of gallic acid and catechin equivalence (GAE: gallic acid equivalent; CE: catechin equivalent) at 750 and 510 nm, respectively.

Three replicates of different concentrations of gallic acid and catechin (10, 20, 30, 40, 50, 100, 150, 200, and 300 μg ml^{-1}) were used to deduce the standard curves for determination of phenolic and flavonoid content, respectively. The generated equation for the gallic acid standard curve was $y = 0.0043x + 0.0019$ $(R^2 = 0.9995)$. Furthermore, the generated one for the catechin standard curve was $y = 0.0034x - 0.0039$ $(R^2 = 0.9993)$. The result obtained from the total phenolic content estimation of the in vitro leaves' extracts was 11.53 mg GAE g^{-1} DW. However, the total flavonoid content was estimated as 12.99 mg CE g^{-1} DW.

From the rich plant sources of phenolics, *Acacia nilotica*, *Acacia catechu*, and *Albizia lebbeck* contain 80.63, 78.12, and 66.23 mg GAE, respectively [72]. Moreover, higher phenolic contents were estimated in the fruits of *Solanum indicum* and *S. surattense* of 250.4–289.5 mg GAE g^{-1} DW [73]. Despite the relatively lower total phenolics detected in this study, the global problem of food shortage necessitates the search for nutritional alternatives as well as nutritional supplements that preserve human health and vitality. On the other hand, the production of the active substance in vitro will remain the most appropriate solution that saves time and effort, especially due to the decline of global cultivated land and climate risks.

3.4. HPTLC Analysis

During the estimation of ferulic acid in dry leaves' extract, the retardation factor (Rf) of the 400 μg ml^{-1} standard was 0.62 (Figure 6). The eight reference volumes (2–9 μL) of the standard were used to generate a linear calibration curve. The linear equation obtained was $y = 5.601 \times 10^{-8}x$ *where* $R = 95.21\%$, and the coefficient of variation (CV) was 11.77%. Only four of the seven different volumes of dry leaves' extract samples (2, 4, 6, and 8 μL) were detected in the calibration range (Figure S5). The final results showed that the mean of ferulic acid content in the three samples within the calibration range was 45.52 mg g^{-1} DW where the CV = 1.19% (Table 4). The HPTLC method was simple, reproducible, and sensitive in the separation and determination of ferulic acid. It was used to estimate ferulic acid in *Lycopodium clavatum* [74], *Setaria italica* [75], and *Ricinus communis* Linn. [76].

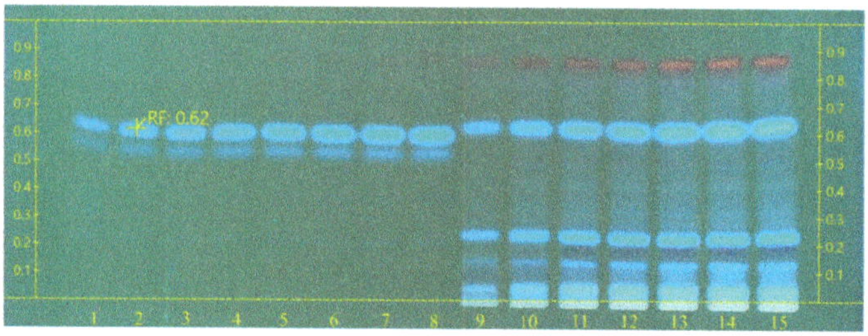

Figure 6. HPTLC chromatogram of *L. schweinfurthii* micropropagated leaves' extract against ferulic acid standard captured at 366 nm. Tracks 1–8: ferulic acid, 400 μg ml^{-1} of volume 2–9 μL; Tracks 9–15: dry leaves' extract of volume 2, 4, 6, 8, 10, 12, and 14 μL.

Table 4. A summary of the results of the total phenolic content, total flavonoid content, ferulic acid content, and antioxidant activity of *micropropagated* L. schweinfurthii dried leaves.

Contents and Antioxidant Capacity	Obtained Results
Total phenolic content	11.53 GAE g^{-1} DW
Total flavonoid content	12.99 CE g^{-1} DW
Ferulic acid content	45.52 mg g^{-1} DW
IC_{50} with DPPH analysis	0.43 mg mL^{-1}
IC_{50} with $ABTS^+$ analysis	1.99 mg mL^{-1}

3.5. Antioxidant Activities

The results obtained from the antioxidant assay revealed that 0.43 mg mL^{-1} of the in vitro leaves' extract were required to scavenge half of the DPPH stable radicals (IC_{50}). However, 1.99 mg mL^{-1} of the leaves' extract were required to scavenge half of the stable ABTS free radicals (Table 4). According to plotting the inhibitory effect, the sensitivity and efficiency of the DPPH assay were higher than the ABTS assay. On the other hand, only 107.57 and 94.71 µg ml^{-1} of black pepper extracts were required to scavenge half of the DPPH and ABTS stable radicals, respectively [55].

4. Conclusions

In this study, we successfully established a suitable, rapid, and efficient protocol for in vitro micropropagation of *L. schweinfurthii* from nodal segments. Reproducible genetic and biochemical techniques were performed to determine the stability of plant genome and expressed proteins in regenerated in vitro plants. The importance of the leaves' extract was proven through the content and activity. This protocol should be useful in future studies for in vitro secondary metabolite production from this plant.

Supplementary Materials: The following are available online at https://www.mdpi.com/article/10.3390/plants10102089/s1, Figure S1: Jazirat Al-Kawm Al-Akhdar (the green islet) which is located in Burullus Lake (northern of Nile Delta) in Egypt showing the populations of *Lycium schweinfurthii*. Figure S2: The blooming of *L. schweinfurthii* during the spring season. Figure S3: *L.schweinfurthii* branch showing leaves and immature fruits. Figure S4: Ripe fruits of *L. schweinfurthii*. Figure S5: Calibration range of the HPTLC analysis of micropropagated dry leaves' extract samples against a reference of ferulic acid 400 µg ml^{-1}.

Author Contributions: Conceptualization, D.M. and H.A.M.M.; Methodology, D.M., A.M.M.G. and I.S.; Validation, H.A.M.M., E.A.E. and I.S.; Investigation, Funding Acquisition, and Software, D.M. and I.S.; Data Curation, D.M., E.A.E. and I.S.; Statistical Analysis, Writing—Original Draft Preparation, D.M.; Writing—Review and Editing, D.M., H.A.M.M., A.M.M.G. and I.S.; Visualization, A.M.M.G., E.A.E. and I.S.; Resources, Supervision, and Project Administration, I.S. All authors have read and agreed to the published version of the manuscript.

Funding: D.M. was funded by a full scholarship (No. 308923) from the Ministry of Higher Education of the Arab Republic of Egypt. This Article is funded by the Open Access Publication Fund of Weihenstephan-Triesdorf University of Applied Sciences.

Acknowledgments: The authors thank Mahmoud H. Sultan, Ramadan Bedair, and Osama Gamal for their help in collecting the plant material and identification of the plant species.

Conflicts of Interest: The authors declare no conflict of interest.

References

1. Yao, R.; Heinrich, M.; Weckerle, C.S. The Genus Lycium as Food and Medicine: A Botanical, Ethnobotanical and Historical Review. *J. Ethnopharmacol.* **2018**, *212*, 50–66. [CrossRef] [PubMed]
2. Stearn, W.T. Lycium. In *Flora Europaea*; Tutin, T.G., Heywood, V.H., Burges, N.A., Moore, D.M., Valentine, D.H., Walters, S.M., Webb, D.A., Eds.; Cambridge University Press: Cambridge, UK, 1972; p. 194.
3. Shaltout, K.H.; El-Din, A.S.; El-Fahar, R.A.; Beshara, H.M. Associated Species and Threats upon *Lycium Schweinfurthii* var. *Schweinfurthii* in the Deltaic Mediterranean Coast, Egypt. *Taeckholmia* **2018**, *38*, 107–122.
4. El-Amier, Y.; El-Halawany, E.; Abdullah, T. Composition and Diversity of Plant Communities in Sand Formations Along the Northern Coast of the Nile Delta in Egypt. *Res. J. Pharm. Biol. Chem. Sci.* **2014**, *5*, 826–847.
5. Khafagi, A.; El- Ghamery, A.; Ghaly, O.; Ragab, O. Fruit and Seed Morphology of Some Species of Solanaceae. *Taeckholmia* **2018**, *38*, 123–140. [CrossRef]
6. Cardoso, J.C.; Sheng Gerald, L.T.; Teixeira da Silva, J.A. Micropropagation in the Twenty-First Century. *Methods Mol. Biol.* **2018**, *1815*, 17–46. [CrossRef] [PubMed]
7. Cui, Y.; Deng, Y.; Zheng, K.; Hu, X.; Zhu, M.; Deng, X.; Xi, R. An Efficient Micropropagation Protocol for an Endangered Ornamental Tree Species (*Magnolia sirindhorniae* Noot. & Chalermglin) and Assessment of Genetic Uniformity through DNA Markers. *Sci. Rep.* **2019**, *9*, 1–10. [CrossRef]

8. Kikowska, M.; Thiem, B.; Szopa, A.; Ekiert, H. Accumulation of Valuable Secondary Metabolites: Phenolic Acids and Flavonoids in Different in vitro Systems of Shoot Cultures of the Endangered Plant Species—*Eryngium alpinum* L. *Plant Cell Tissue Organ Cult.* **2020**, *141*, 381–391. [CrossRef]
9. Gao, Y.; Wang, Q.-M.; An, Q.; Cui, J.; Zhou, Y.; Qi, X.; Zhang, L.; Li, L. A Novel Micropropagation of *Lycium ruthenicum* and Epigenetic Fidelity Assessment of Three Types of Micropropagated Plants in vitro and ex vitro. *PLoS ONE* **2021**, *16*, e0247666. [CrossRef]
10. Samiei, L.; Mirshahi, H.; Pahnehkolayi, M.D.; Tehranifar, A. Micropropagation of *Lycium depressum*, a Promising Native Shrub for Urban landscape. *Iran. J. Hortic. Sci.* **2020**, *51*, 741–751. [CrossRef]
11. Silvestri, C.; Sabbatini, G.; Marangelli, F.; Rugini, E.; Cristofori, V. Micropropagation and ex vitro Rooting of Wolfberry. *HortScience* **2018**, *53*, 1494–1499. [CrossRef]
12. Jung, W.-S.; Chung, I.-M.; Kim, S.-H.; Chi, H.-Y.; Yu, C.Y.; Ghimire, B.K. Direct Shoot Organogenesis from *Lycium chinense* Miller Leaf Explants and Assessment of Genetic Stability Using ISSR Markers. *Agronomy* **2021**, *11*, 503. [CrossRef]
13. Samaha, G.M.; Ahmed, M.A.; Abd El-Hameid, A.R. Assessment of Growth and Productivity of Five Peanut Cultivars and Genetic Diversity Using RAPD Markers. *Bull. Natl. Res. Cent.* **2019**, *43*, 1–11. [CrossRef]
14. Thakur, M.; Soni, M.; Sharma, D.P.; Vivek, M.; Sharma, V. In vitro Propagation of Plum (*Prunus salicina*) cv. 'Santa Rosa' and Assessment of Genetic Stability Using RAPD Markers. *Indian J. Plant Physiol.* **2018**, *23*, 161–168. [CrossRef]
15. Lema-Rumińska, J.; Kulus, D.; Tymoszuk, A.; Varejão, J.M.T.B.; Bahcevandziev, K. Profile of Secondary Metabolites and Genetic Stability Analysis in New Lines of *Echinacea purpurea* (L.) Moench Micropropagated via Somatic Embryogenesis. *Ind. Crops Prod.* **2019**, *142*, 111851. [CrossRef]
16. Tikendra, L.; Potshangbam, A.M.; Dey, A.; Devi, T.R.; Sahoo, M.R.; Nongdam, P. RAPD, ISSR, and SCoT Markers Based Genetic Stability Assessment of Micropropagated *Dendrobium fimbriatum* Lindl. Var. Oculatum Hk. f.-an Important Endangered Orchid. *Physiol. Mol. Biol. Plants* **2021**, *27*, 341–357. [CrossRef]
17. Oliya, B.K.; Chand, K.; Thakuri, L.S.; Baniya, M.K.; Sah, A.K.; Pant, B. Assessment of Genetic Stability of Micropropagated Plants of *Rhynchostylis retusa* (L.) Using RAPD Markers. *Sci. Hortic.* **2021**, *281*, 110008. [CrossRef]
18. Patel, P.; Rajkumar, B.K.; Parmar, P.; Shah, R.; Krishnamurthy, R. Assessment of Genetic Diversity in *Colletotrichum falcatum* Went Accessions Based on RAPD and ISSR Markers. *J. Genet. Eng. Bio-Technol.* **2018**, *16*, 153–159. [CrossRef]
19. Wójcik, D.; Trzewik, A.; Kucharska, D. Field Performance and Genetic Stability of Micropropagated Gooseberry Plants (*Ribes grossularia* L.). *Agronomy* **2021**, *11*, 45. [CrossRef]
20. Tikendra, L.; Amom, T.; Nongdam, P. Molecular Genetic Homogeneity Assessment of Micropropagated *Dendrobium moschatum* Sw.—A Rare Medicinal Orchid, Using RAPD and ISSR Markers. *Plant Gene* **2019**, *19*, 100196. [CrossRef]
21. Chittora, M. Assessment of Genetic Fidelity of Long Term Micropropagated Shoot Cultures of *Achras sapota* L. Var. "Cricket Ball" as Assessed by RAPD and ISSR Markers. *Indian J. Biotechnol.* **2018**, *17*, 492–495.
22. Ahmed, M.R.; Anis, M.; Alatar, A.A.; Faisal, M. In Vitro Clonal Propagation and Evaluation of Genetic Fidelity Using RAPD and ISSR Marker in Micropropagated Plants of *Cassia alata* L.: A Potential Medicinal Plant. *Agrofor. Syst.* **2017**, *91*, 637–647. [CrossRef]
23. Jong, L.W.; Thien, V.Y.; Yong, Y.S.; Rodrigues, K.F.; Yong, W.T.L. Micropropagation and Protein Profile Analysis by SDS-PAGE of *Gracilaria changii* (Rhodophyta, Solieriaceae). *Aquac. Rep.* **2015**, *1*, 10–14. [CrossRef]
24. Khattab, S.; El Sherif, F.; El-Garhy, H.A.; Ahmed, S.; Ibrahim, A. Genetic and Phytochemical Analysis of the in vitro Regenerated *Pilosocereus robinii* by ISSR, SDS-PAGE and HPLC. *Gene* **2014**, *533*, 313–321. [CrossRef]
25. Mahmoud, R.A.; Hassan, O.S.; Abou-Hashish, A.; Amin, A. Role of Trehalose during Recovery from Drought Stress in Micropropagated Banana (*Musa* Spp.) Transplants. *Res. J. Pharm. Biol. Chem. Sci.* **2017**, *8*, 1335–1345.
26. El-Mageid, I.S. Evaluation of Genetic Stability by Using Protein and ISSR Markers during Callus Development Stage of Some Date Palm (*Phoenix dactylifera* L.) Cultivars under Effect of 2,4-D and Picloram. *Middle East J. Appl. Sci.* **2019**, *9*, 483–493.
27. Konarska, A. Microstructural and Histochemical Characteristics of *Lycium barbarum* L. Fruits Used in Folk Herbal Medicine and as Functional Food. *Protoplasma* **2018**, *255*, 1839–1854. [CrossRef]
28. Wink, M. Modes of Action of Herbal Medicines and Plant Secondary Metabolites. *Medicines* **2015**, *2*, 251–286. [CrossRef] [PubMed]
29. Ewais, E.A.; Abd El-Maboud, M.M.; Elhaw, M.H.; Haggag, M.I. Phytochemical Studies on *Lycium schweinfurthii* Var. Schweinfurthii (Solanaceae) and Isolation of Five Flavonoids from Leaves. *J. Med. Plants Stud.* **2016**, *4*, 288–300.
30. Cushnie, T.P.T.; Lamb, A.J. Recent Advances in Understanding the Antibacterial Properties of Flavonoids. *Int. J. Antimicrob. Agents* **2011**, *38*, 99–107. [CrossRef]
31. Tattini, M.; Galardi, C.; Pinelli, P.; Massai, R.; Remorini, D.; Agati, G. Differential Accumulation of Flavonoids and Hydroxycinnamates in Leaves of *Ligustrum vulgare* under Excess Light and Drought Stress. *New Phytol.* **2004**, *163*, 547–561. [CrossRef] [PubMed]
32. Barceló, J.; Poschenrieder, C. Fast Root Growth Responses, Root Exudates, and Internal Detoxification as Clues to the Mechanisms of Aluminium Toxicity and Resistance: A Review. *Environ. Exp. Bot.* **2002**, *48*, 75–92. [CrossRef]
33. Ryan, K.G.; Swinny, E.E.; Markham, K.R.; Winefield, C. Flavonoid Gene Expression and UV Photoprotection in Transgenic and Mutant *Petunia* Leaves. *Phytochemistry* **2002**, *59*, 23–32. [CrossRef]
34. Treutter, D. Significance of Flavonoids in Plant Resistance: A Review. *Environ. Chem. Lett.* **2006**, *4*, 147–157. [CrossRef]
35. Lin, D.; Xiao, M.; Zhao, J.; Li, Z.; Xing, B.; Li, X.; Kong, M.; Li, L.; Zhang, Q.; Liu, Y.; et al. An Overview of Plant Phenolic Compounds and Their Importance in Human Nutrition and Management of Type 2 Diabetes. *Molecules* **2016**, *21*, 1374. [CrossRef]

36. Pietta, P.G. Flavonoids as Antioxidants. *J. Nat. Prod.* **2000**, *63*, 1035–1042. [CrossRef]
37. Maleki, S.J.; Crespo, J.F.; Cabanillas, B. Anti-Inflammatory Effects of Flavonoids. *Food Chem.* **2019**, *299*, 125124. [CrossRef]
38. Yamamoto, Y.; Gaynor, R.B. Therapeutic Potential of Inhibition of the NF-κB Pathway in the Treatment of Inflammation and Cancer. *J. Clin. Invest.* **2001**, *107*, 135–142. [CrossRef]
39. Lani, R.; Hassandarvish, P.; Shu, M.H.; Phoon, W.H.; Chu, J.J.H.; Higgs, S.; Vanlandingham, D.; Abu Bakar, S.; Zandi, K. Antiviral Activity of Selected Flavonoids against Chikungunya Virus. *Antiviral Res.* **2016**, *133*, 50–61. [CrossRef] [PubMed]
40. Manner, S.; Skogman, M.; Goeres, D.; Vuorela, P.; Fallarero, A. Systematic Exploration of Natural and Synthetic Flavonoids for the Inhibition of *Staphylococcus aureus* Biofilms. *Int. J. Mol. Sci.* **2013**, *14*, 19434–19451. [CrossRef] [PubMed]
41. Peterson, J.J.; Dwyer, J.T.; Jacques, P.F.; McCullough, M.L. Associations between Flavonoids and Cardiovascular Disease Incidence or Mortality in European and US Populations. *Nutr. Rev.* **2012**, *70*, 491–508. [CrossRef] [PubMed]
42. Rezende, B.A.; Pereira, A.; Cortes, S.; Lemos, V. Vascular Effects of Flavonoids. *Curr. Med. Chem.* **2016**, *23*, 87–102. [CrossRef]
43. Cazarolli, L.; Zanatta, L.; Alberton, E.; Bonorino Figueiredo, M.S.; Folador, P.; Damazio, R.; Pizzo-latti, M.; Barreto Silva, F.R. Flavonoids: Prospective Drug Candidates. *Mini-Rev. Med. Chem.* **2008**, *8*, 1429–1440. [CrossRef] [PubMed]
44. Jamous, R.; Zaitoun, S.; Husein, A.; Qasem, I.; Ali-Shtayeh, M. Screening for Biological Activities of Medicinal Plants Used in Traditional Arabic Palestinian Herbal Medicine. *Eur. J. Med. Plants* **2015**, *9*, 1–13. [CrossRef]
45. Elbermawi, A.; Halim, A.F.; Mansour, E.S.S.; Ahmad, K.F.; Ashour, A.; Amen, Y.; Shimizu, K. A New Glucoside with a Potent α-Glucosidase Inhibitory Activity from *Lycium schweinfurthii*. *Nat. Prod. Res.* **2021**, *35*, 976–983. [CrossRef]
46. Smetanska, I. Sustainable Production of Polyphenols and Antioxidants by Plant in vitro Cultures. In *Bioprocessing of Plant In Vitro Systems*; Pavlov, A., Bley, T., Eds.; Springer: Cham, Switzerland, 2018; pp. 225–269. ISBN 978-3-319-54599-8.
47. Ali, A.M.A.; El-Nour, M.E.A.M.; Yagi, S.M. Total Phenolic and Flavonoid Contents and Antioxidant Activity of Ginger (*Zingiber officinale* Rosc.) Rhizome, Callus and Callus Treated with Some Elicitors. *J. Genet. Eng. Biotechnol.* **2018**, *16*, 677–682. [CrossRef]
48. El-Hawary, S.S.; Abd El-Kader, E.M.; Rabeh, M.A.; Abdel Jaleel, G.A.; Arafat, M.A.; Schirmeister, T.; Abdelmohsen, U.R. Eliciting Callus Culture for Production of Hepatoprotective Flavonoids and Phenolics from *Sequoia sempervirens* (D. Don Endl). *Nat. Prod. Res.* **2020**, *34*, 3125–3129. [CrossRef]
49. Darwish, H.Y.; Ahmed, S.M. Elicitors Enhancing Phenolics Content and Related Gene Expression Variation in Petal—Derived Calli of *Rosa damascena* Mill. *Egypt. J. Bot.* **2020**, *60*, 71–79. [CrossRef]
50. Çelik, M.; Keskin, N.; Özdemir, F.A. The Effects of UV Irradiation and Incubation Time on in vitro Phenolic Compound Production in "Karaerik" Callus Culture. *Kahramanmaraş Sütçü İmam Üniversitesi Tarım Ve Doğa Derg.* **2020**, *23*, 1428–1434. [CrossRef]
51. Martins, M.; Tenreiro, R.; Oliveira, M.M. Genetic Relatedness of Portuguese Almond Cultivars Assessed by RAPD and ISSR Markers. *Plant Cell Rep.* **2003**, *22*, 71–78. [CrossRef] [PubMed]
52. Williams, J.G.K.; Kubelik, A.R.; Livak, K.J.; Rafalski, J.A.; Tingey, S.V. DNA Polymorphisms Amplified by Arbitrary Primers Are Useful as Genetic Markers. *Nucleic Acids Res.* **1990**, *18*, 6531–6535. [CrossRef]
53. Laemmli, U.K. Cleavage of Structural Proteins during the Assembly of the Head of Bacteriophage T4. *Nature* **1970**, *227*, 680–685. [CrossRef] [PubMed]
54. Marinova, D.; Ribarova, F.; Atanassova, M. Total Phenolics and Total Flavonoids in Bulgarian Fruits and Vegetables. *J. Univ. Chem. Technol. Metall.* **2005**, *40*, 255–260.
55. Olalere, O.A.; Abdurahman, H.N.; Yunus, R.B.M.; Alara, O.R.; Ahmad, M.M.; Zaki, Y.H.; Abdlrhman, H.S.M. Parameter Study, Antioxidant Activities, Morphological and Functional Characteristics in Microwave Extraction of Medicinal Oleoresins from Black and White Pepper. *J. Taibah Univ. Sci.* **2018**, *12*, 730–737. [CrossRef]
56. Gabr, A.M.M.; Mabrok, H.B.; Ghanem, K.Z.; Blaut, M.; Smetanska, I. Lignan Accumulation in Callus and *Agrobacterium Rhizogenes*-Mediated Hairy Root Cultures of Flax (*Linum usitatissimum*). *Plant Cell Tissue Organ Cult.* **2016**, *126*, 255–267. [CrossRef]
57. Sneath, P.H.A.; Sokal, R.R. *Numerical Taxonomy: The Principles and Practice of Numerical Classification*; W. H. Freeman: San Francisco, CA, USA, 1973.
58. Hasan, M.N.; Nigar, S.; Rabbi, M.A.K.; Mizan, S.B.; Rahman, M.S. Micropropagation Of Strawberry (*Fragaria x Ananassa* Duch.). *Int. J. Sustain. Crop. Prod.* **2010**, *5*, 36–41.
59. Amghar, I.; Diria, G.; Boumlik, I.; Gaboun, F.; Iraqi, D.; Labhilili, M.; Mentag, R.; Meziani, R.; Mazri, M.A.; Ibriz, M.; et al. An Efficient Regeneration Pathway through Adventitious Organogenesis for the Endangered *Argania spinosa* (L.) Skeels. *Vegetos* **2021**, *34*, 355–367. [CrossRef]
60. Thomas, T.D.; Maseena, E.A. Callus Induction and Plant Regeneration in *Cardiospermum halicacabum* Linn. an Important Medicinal Plant. *Sci. Hortic.* **2006**, *108*, 332–336. [CrossRef]
61. Irvani, N.; Solouki, M.; Omidi, M.; Zare, A.R.; Shahnazi, S. Callus Induction and Plant Regeneration in *Dorem ammoniacum* D., an Endangered Medicinal Plant. *Plant Cell Tissue Organ Cult.* **2010**, *100*, 293–299. [CrossRef]
62. Sen, M.K.; Nasrin, S.; Rahman, S.; Jamal, A.H.M. In Vitro Callus Induction and Plantlet Regeneration of *Achyranthes aspera* L., a High Value Medicinal Plant. *Asian Pac. J. Trop. Biomed.* **2014**, *4*, 40–46. [CrossRef]
63. Ozdemir, F.A.; Gur, N. In Vitro Propagation of Cataloglu Apricot (*Prunus armeniaca* L.) Cultivar Using Apical Node As Explant. *Prog. Nutr.* **2018**, *20*, 176–181. [CrossRef]
64. El-Ghamery, A.A.; Khafagi, A.A.F.; Ragab, O.G. Taxonomic Implication of Pollen Morphology and Seed Protein Electrophoresis of Some Species of Solanaceae in Egypt. *Azhar Bull. Sci.* **2018**, *29*, 43–54.

65. Khawale, R.N.; Singh, S.K.; Yerramilli, V.; Grover, M. Assessment of Clonal Fidelity of Micropropagated Grape (*Vitis vinifera* L.) Plants by RAPD Analysis. *Physiol. Mol. Biol. Plants* **2006**, *12*, 189–192.
66. Lakshmanan, V.; Reddampalli Venkataramareddy, S.; Neelwarne, B. Molecular Analysis of Genetic Stability in Long-Term Micropropagated Shoots of Banana Using RAPD and ISSR Markers. *Electron. J. Biotechnol.* **2007**, *10*, 717–3458. [CrossRef]
67. Martins, M.; Sarmento, D.; Oliveira, M.M. Genetic Stability of Micropropagated Almond Plantlets, as Assessed by RAPD and ISSR Markers. *Plant Cell Rep.* **2004**, *23*, 492–496. [CrossRef] [PubMed]
68. Parida, R.; Mohanty, S.; Nayak, S. Molecular Characterization of Endangered Medicinal Plant Species *Hedychium coronarium* from Eastern India. *Int. J. Pharm. Pharm. Sci.* **2017**, *9*, 173–178. [CrossRef]
69. Osman, G.; Munshi, A.; Altf, F.; Mutawie, H. Genetic Variation and Relationships of *Zea mays* and *Sorghum* Species Using RAPD-PCR and SDS-PAGE of Seed Proteins. *Afr. J. Biotechnol.* **2013**, *12*, 4269–4276. [CrossRef]
70. Seesangboon, A.; Pokawattana, T.; Eungwanichayapant, P.D.; Tovaranonte, J.; Popluechai, S. Effects of 6-Benzyladenine on *Jatropha* Gene Expression and Flower Development. *Russ. J. Plant Physiol.* **2018**, *65*, 345–356. [CrossRef]
71. Alizadeh, M.; Singh, S.K. Molecular Assessment of Clonal Fidelity in Micropropagated Grape (*Vitis* Spp.) Rootstock Genotypes Using RAPD and ISSR Markers. *Iran. J. Biotechnol.* **2009**, *7*, 17–44.
72. Sulaiman, C.T.; Balachandran, I. Total phenolics and total flavonoids in selected Indian medicinal plants. *Indian J. Pharm. Sci.* **2012**, *74*, 258–260. [CrossRef] [PubMed]
73. Yasir, M.; Sultana, B.; Anwar, F. LC–ESI–MS/MS Based Characterization of Phenolic Components in Fruits of Two Species of Solanaceae. *J. Food Sci. Technol.* **2018**, *55*, 2370–2376. [CrossRef] [PubMed]
74. Srivastava, S.; Singh, A.P.; Singh Rawat, A.K. A HPTLC Method for the Identification of Ferulic Acid from *Lycopodium clavatum*. *Asian Pac. J. Trop. Biomed.* **2012**, *2*, S12–S14. [CrossRef]
75. Goudar, G.; Sathisha, G.J. Effect of Processing on Ferulic Acid Content in Foxtail Millet (*Setaria italica*) Grain Cultivars Evaluated by HPTLC. *Orient. J. Chem.* **2016**, *32*, 2251–2258. [CrossRef]
76. Verma, S.C.; Rani, R.; Pant, P.; Padhi, M.M.; Jain, C.L.; Babu, R. Quantitative Determination of Ferulic Acid in *Ricinus communis* Linn. Leaves and Its Geographical Variation Using HPTLC Finger-print. *Chem. Sin.* **2011**, *2*, 127–135.

plants

Article

Optimization of Biomass Accumulation and Production of Phenolic Compounds in Callus Cultures of *Rhodiola rosea* L. Using Design of Experiments

Anna A. Erst [1,*], Anastasia A. Petruk [1], Andrey S. Erst [1,2], Denis A. Krivenko [3], Nadezhda V. Filinova [3], Svetlana Y. Maltseva [4], Maxim S. Kulikovskiy [4] and Evgeny V. Banaev [1]

1 Central Siberian Botanical Garden, Siberian Branch of Russian Academy of Sciences, 630090 Novosibirsk, Russia; pet.a@mail.ru (A.A.P.); erst_andrew@yahoo.com (A.S.E.); alnus2005@mail.ru (E.V.B.)

2 Laboratory of Plants Systematics and Phylogeny, National Research Tomsk State University, 634050 Tomsk, Russia

3 Siberian Institute of Plant Physiology & Biochemistry, Siberian Branch of Russian Academy of Sciences, 664033 Irkutsk, Russia; krivenko.irk@gmail.com (D.A.K.); filinova_nv@mail.ru (N.V.F.)

4 K.A. Timiryazev Institute of Plant Physiology, Russian Academy of Sciences, 127276 Moscow, Russia; svetadm32@gmail.com (S.Y.M.); max-kulikovsky@yandex.ru (M.S.K.)

* Correspondence: erst_aa@csbg-nsk.ru

Abstract: *Rhodiola rosea* L. is a valuable medicinal plant with adaptogenic, neuroprotective, antitumor, cardioprotective, and antidepressant effects. In this study, design of experiments methodology was employed to analyze and optimize the interacting effects of mineral compounds (concentration of NO_3^- and the ratio of NH_4^+ to K^+) and two plant growth regulators [total 6-benzylaminopurine (BAP) and α-naphthylacetic acid (NAA) concentration and the ratio of BAP to NAA] on the growth and the production of total phenolic compounds (TPCs) in *R. rosea* calluses. The overall effect of the model was highly significant ($p < 0.0001$), indicating that NH_4^+, K^+, NO_3^-, BAP, and NAA significantly affected growth. The best callus growth (703%) and the highest production of TPCs (75.17 mg/g) were achieved at an NH_4^+/K^+ ratio of 0.33 and BAP/NAA of 0.33, provided that the concentration of plant growth regulators was 30 μM and that of NO_3^- was ≤40 mM. According to high-performance liquid chromatography analyses of aerial parts (leaves and stems), in vitro seedlings and callus cultures of *R. rosea* contain no detectable rosarin, rosavin, rosin, and cinnamyl alcohol. This is the first report on the creation of an experiment for the significant improvement of biomass accumulation and TPC production in callus cultures of *R. rosea*.

Keywords: roseroot; in vitro culture; design of experiments; nitrogen source; plant growth regulator; methyl jasmonate; HPLC; phenolic compound; histochemistry

Citation: Erst, A.A.; Petruk, A.A.; Erst, A.S.; Krivenko, D.A.; Filinova, N.V.; Maltseva, S.Y.; Kulikovskiy, M.S.; Banaev, E.V. Optimization of Biomass Accumulation and Production of Phenolic Compounds in Callus Cultures of *Rhodiola rosea* L. Using Design of Experiments. *Plants* **2022**, *11*, 124. https://doi.org/10.3390/plants11010124

Academic Editors: Laura Pistelli and Kalina Danova

Received: 30 November 2021
Accepted: 30 December 2021
Published: 2 January 2022

1. Introduction

The application of plant in vitro systems as a sustainable platform for the biotechnological production of pharmaceuticals is a promising alternative to the traditional pipeline. In vitro systems possess numerous advantages, including biosynthesis of safe metabolites according to good manufacturing practices (GMP) and independence from environmental factors [1–3]. In addition, this approach does not threaten natural populations of rare and endangered plant species. One of the species currently in demand for the biotechnological production of natural substances, having adaptogenic properties and various medicinal effects [4–6], is *Rhodiola rosea* L. from the Crassulacea family. Recently, *R. rosea* was actively used in the manufacture of various dietary supplements [7]; large volumes of harvesting in the wild and a slow rate of renewal have put this species on the brink of extinction. Notably, most of the products on the market are based on the raw material collected from wild populations in the Altai region (Russia) [5]. It is fairly well documented that the standard in vitro culture of *R. rosea* is not efficient enough to compete with wild plants with respect to the accumulation of active ingredients. In almost all earlier reports, researchers admit the lack or only traces of the important compounds in their in vitro experiments with roseroot [8–11]. In in vitro cultures, a significant enhancement of the production of rosin and its derivatives is observed when the cultures are fed with a precursor: cinnamyl alcohol [2,8,12]. In *R. rosea* compact callus aggregate cultures, the observed rosin and rosarin content is even higher than that in field-cultivated plants, while the rosavin level is five times lower. Recently developed hairy root cultures of *Rhodiola kirilowii* (Regel) Maxim. supplemented with cinnamyl alcohol exhibit a higher potential for the production of rosin and its derivatives in comparison to field-cultivated plants [13]. It has been shown that light quality has a stimulatory effect on secondary-metabolite production in callus cultures of *R. imbricate* Edgew [14]. Therefore, new approaches should be developed to overcome the shortage of active ingredients in in vitro cultures.

In this study, the design of experiments (DoE) methodology was utilized to examine the parameters that affect callus biomass accumulation and production of phenolic compounds by *R. rosea*. DoE is a statistical methodology that allows simultaneous testing of multiple factors to understand and improve complex systems [15,16].

Mineral nutrients are some of the most basic components of plant tissue culture media. Nitrogen (N) in the form of NH_4^+ or NO_3^- is the dominant mineral nutrient in most tissue culture formulations [17]. The culture of isolated plant tissues is autotrophic with respect to the N source. From inorganic N sources, tissues synthesize all organic nitrogenous compounds necessary for normal physiological processes [18]. The concentration and form of N in tissue culture media have a significant influence on cell growth and differentiation [19]. The most common N forms used in tissue culture growth media are NO_3^- and NH_4^+. The effects of N may be dependent on either the total N concentration or the proportion of NO_3^- and NH_4^+. In general, NO_3^- is the favored form for N assimilation in most plants; NH_4^+ is sometimes not required and at high concentrations may be toxic [19,20]. For most plants, a combination of NO_3^- and NH_4^+ is better than either NO_3^- or NH_4^+ as a sole source of N. Changes of NH_4NO_3 and KNO_3 concentrations will alter the concentrations and proportions of K^+, NO_3^-, and NH_4^+ in the culture medium [21]. In plant tissue culture media, N is present as various salts and varying the proportions of salts creates the problem of ion confounding between the effect of the N source and the effect of the counter ion in that salt. Varying the NH_4^+/K^+ ratio and the total nitrate ion concentration NO_3^- in a two-component mixture facilitates the direct estimation of mineral nutrients' effects without the ion confounding of a salt-based approach [17]. Computer-aided experimental design helps formulate practical treatments consisting of several factors or mixtures for studying the effects of complicated systems, in contrast to the traditional studies or factorial designs [22].

Various concentrations and combinations of plant growth regulators have been used to obtain callus and suspension cultures of *Rhodiola* species. For the induction of callus formation in these species, a combination of cytokinins and auxins is required [10,23–31]. 6-benzylaminopurine (BAP) is one of the most widely used plant growth regulators for in vitro culture of *Rhodiola*. Cotyledons, hypocotyl, leaves, apical buds, and internode fragments, inoculated on media containing BAP (0.2–3 mg/L) in combination with indole-3-acetic acid (IAA; 0.1 mg/L) [24], α-naphthylacetic acid (NAA; 0.5–3 mg/L) [10,28], or 2,4-dichlorophenoxyacetic acid (2,4-D; 0.5–3.0 mg/L) [25,31] have been the most appropriate explant types and media constituents for induction of well-growing calluses [5].

Numerous authors have demonstrated that the addition of an elicitor (biotic or abiotic) to a culture medium significantly increases the production of secondary metabolites in vitro by triggering a metabolic cascade [32,33]. Jasmonic acid and methyl jasmonate (MJ) are recognized as effective elicitors that trigger the biosynthesis of secondary defense compounds in callus and suspension cultures by activating the genes of secondary metabolism [34]. For example, jasmonic acid has been shown to enhance phenolic-compound production and bioactivity in a suspension culture of *R. imbricate* [28].

In the present study, the DoE approach was used to explore the relations among multiple factors in a new rapid culture method and their influence on process outcomes. DoE is particularly useful for examining interactions among factors that cannot be predicted by experiments designed to test one factor at a time (OFAT) [35]. In our study, four factors were tested that influence biomass accumulation and the production of phenolic compounds in the callus culture of *R. rosea*: ratio NH_4^+/K^+, NO_3^- concentration, ratio BAP/NAA, and BAP + NAA concentration. In addition, in this study, we evaluated the impact of MJ in callus culture under optimal nutrient conditions.

2. Results

A summary of ANOVA (analysis of variance), the lack-of-fit test, and three R^2 statistics for % fresh weight increase and dry weight is presented in Table 1. The percent fresh weight increase in the biomass of *R. rosea* callus after 45 days of cultivation varied from 14% to 703%. The result of the lack-of-fit test was not significant ($p = 0.8180$), indicating that additional variation in the residuals could not be removed with a better model. R^2, adjusted R^2, and predicted R^2 ranged from 0.92 to 0.98. The effect of the overall model was highly significant ($p < 0.0001$), indicating that NH_4^+, K^+, NO_3^-, BAP, and NAA significantly affected the growth. ANOVA revealed 17 significant terms, 12 of which had *p*-values < 0.0001. Dry weight accumulation ranged from 0.05 to 0.26 g. R^2, adjusted R^2, and predicted R^2 ranged from 0.73 to 0.91, indicating good agreement among these five values. The overall effect of the model was highly significant ($p < 0.0001$), indicating significant factor effects on dry weight by these three ions and plant growth regulators. ANOVA revealed nine significant terms, three of which had *p*-values < 0.0001 (Table 1).

2.1. Effects of the NH_4^+/K^+ Ratio and NO_3^- Concentration on Callus Culture

Callus growth in the control treatment group, with the NH_4^+/K^+ ratio at 1.0 and 40 mM NO_3^-, was moderate (approximately average; Figure 1A,B). The highest values were obtained with the NH_4^+/K^+ ratio of 0.33 and 20–40 mM NO_3^-. The influence of the Linear x Linear Mixture component was highly significant for fresh and dry weight. The Linear x Linear Mixture component compares the responses at the extreme ends (vertices) of the mixture design space. This means that growth at the points corresponding to the NH_4^+/K^+ ratio of 0/1 was compared to growth at the points corresponding to the NH_4^+/K^+ ratio of 0.5/0.5. We noted that good growth of callus culture requires $NH_4^+ > 0$. The constructed model showed that the concentration of NO_3^- has a smaller effect on the growth of fresh and dry biomass in comparison with the NH_4^+/K^+ ratio.

2.2. Effects of Ratios BAP/NAA and NH_4^+/K^+ on Callus Culture

The impact of the interaction between ratios BAP/NAA and NH_4^+/K^+ was very highly significant toward the % fresh weight increase and dry weight of callus culture of *R. rosea*. The largest values were achieved at the NH_4^+/K^+ ratio of 0.33 and BAP/NAA 0.33–1.00, provided that the concentration of plant growth regulators was 30 μM and the content of NO_3^- was not more than 40 mM (Figure 1C,D). A change in the parameters of the model showed that with a decrease in the content of NO_3^-, there is a shift of the maximum toward the NH_4^+/K^+ ratio of 1.0. Good growth of callus culture required BAP > 0.

2.3. Effects of BAP + NAA and NO_3^- Concentrations on Callus Culture

The best growth was achieved at high concentrations of plant growth regulators (30 μM) and NO_3^- concentrations of 20–40 mM (Figure 1E,F). We also noted a large increase in dry biomass at a low concentration of sum BAP (5 μM) and NAA (5 μM). The constructed model indicated that peak values were not reached in the experiment.

Table 1. ANOVA and summary statistics for the % fresh weight increase and dry weights of callus culture of *R. rosea*.

	Fresh Weight		Dry Weight	
	F-Value	*p*-Value	F-Value	*p*-Value
Model	107.14	<0.0001	35.83	<0.0001
Linear × Linear Mixture	274.28	<0.0001	80.74	<0.0001
NH_4^+ * K^+ *BAP	140.46	<0.0001	31.82	<0.0001
NH_4^+ * K^+ *NAA	22.00	<0.0001	13.40	0.0008
NH_4^+ *BAP*NAA	38.42	<0.0001	6.61	0.0142
NH_4^+ *BAP* NO_3^-	5.78	0.0218	1.01	0.3208
NH_4^+ *BAP*BAP + NAA	21.64	<0.0001	6.87	0.0126
NH_4^+ * K^+ *BAP* NO_3^-	19.65	<0.0001	-	-
NH_4^+ * K^+ *BAP* BAP + NAA	19.84	<0.0001	2.74	0.1058
NH_4^+ * K^+ *NAA* NO_3^-	7.92	0.0081	1.70	0.2002
NH_4^+ * K^+ *NAA* BAP + NAA	15.35	0.0004	6.63	0.0141
NH_4^+ *BAP*NAA* NO_3^-	76.36	<0.0001	17.88	0.0001
NH_4^+ *BAP*NAA* BAP + NAA	68.45	<0.0001	-	-
NH_4^+ *BAP* NO_3^-* BAP + NAA	21.51	<0.0001	25.17	<0.0001
NH_4^+ * NAA* NO_3^-* BAP + NAA	12.34	0.0013	-	-
K^+ *BAP* $[NO_3^-]^2$	1.84	0.1842	-	-
NH_4^+ * K^+ *BAP* NO_3^-* BAP + NAA	4.74	0.0364	7.54	0.0092
NH_4^+ *BAP*NAA* NO_3^-* BAP + NAA	23.44	<0.0001	1.83	0.1840
NH_4^+ * K^+ *BAP* $[NO_3^-]^2$	61.34	<0.0001	16.24	0.0003
NH_4^+ * K^+ *BAP*[BAP + NAA]^2	7.57	0.0095	3.79	0.0590
Lack of Fit	$p = 0.8180$		$p = 0.6741$	
R^2	0.9851		0.9413	
Adjusted R^2	0.9759		0.9150	
Predicted R^2	0.9235		0.8162	
Adeq Precision	40.4608		23.0849	
Std. Dev.	25.21		0.0142	
Mean	178.00		0.1071	
C.V.%	14.16		13.26	
Model type	Reduced Quadratic × Quadratic × Quadratic model		Reduced Quadratic × Quadratic × Quadratic model	

Note: "-": no hierarchical relationships.

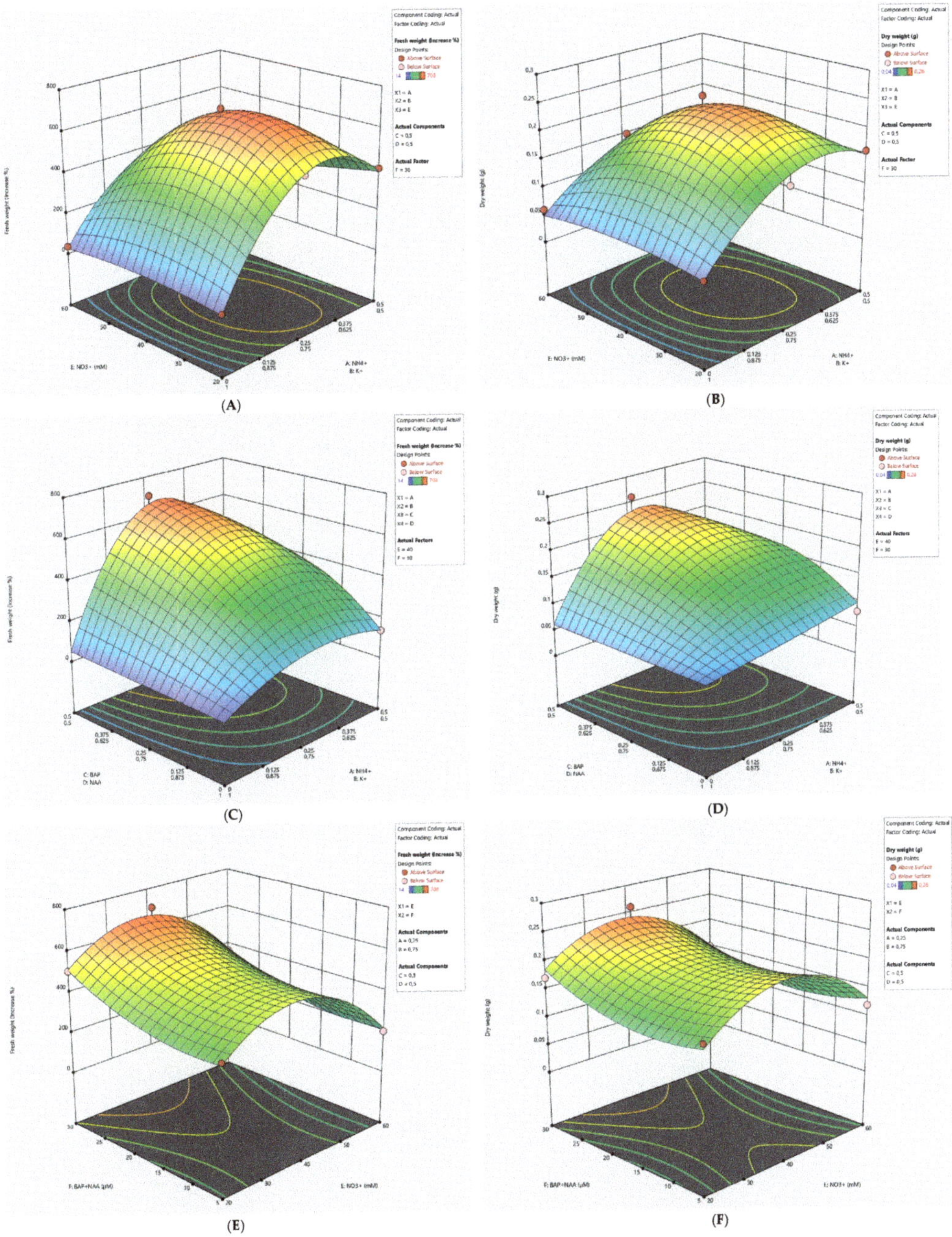

Figure 1. Three-dimensional (3D) response surface plots showing the effects of various factors on *R. rosea* callus fresh and dry weight. (**A**,**B**) The impact of the NH_4^+/K^+ ratio and NO_3^-. (**C**,**D**) The influence of the BAP/NAA ratio and NH_4^+/K^+. (**E**,**F**) The effect of the concentrations of BAP + NAA and NO_3^-.

2.4. Histochemical Analysis

The callus culture obtained under optimal culture conditions was semi-friable yellowish green. Many of the large calluses contained cavities at their centers. Histological analysis revealed that the calluses consisted of actively dividing cells with two types of vacuoles: numerous small vacuoles or a single central one (Figure 2A–D). Small vacuoles are characteristic of immature cells and large vacuoles of mature cells. Lugol's staining showed some accumulation patterns of starch granules in the callus culture (Figure 2E). Starch granules were previously found by us only in the cells of the root and rhizome, while leaves and stems had no starch-containing cells [36]. Cells of the 45-day-old callus did not yield positive signals in a ferric chloride reaction for phenolic compounds. We previously demonstrated that roots and rhizomes tested positive for phenolic compounds with this reagent [36]. In the present study, we found chaotic xylem elements in compacted parts of the callus cultures (Figure 2F).

2.5. Biochemical Analysis

2.5.1. The Profile and Levels of Phenolic Compounds in Callus Culture and Seedlings (Control 1)

Biochemical analysis was carried out only for well-growing callus cultures (an increase in fresh biomass of more than 300%) and for callus grown on the standard MS medium. High-performance liquid chromatography (HPLC; suitable for isolating a complex set of phenolic compounds) detected 27 phenolic compounds in the callus culture and nine phenolic compounds in seedlings of *R. rosea* (control 1) (Table S1). Total phenolic compound (TPC) concentration varied among callus cultures from 14.9 to 71.6 mg/g. The number of phenolic compounds varied from 10 to 20; in control 1, these values were 6.5 mg/g and 9, respectively (Figure 3).

The highest content of TPC was observed in calluses grown on media with the NH_4^+/K^+ ratio of 0.33 and BAP/NAA of 0.33 and a concentration of plant growth regulators of 30 μM (56.01–75.17 mg/g), which is 1.1–1.5 times higher than that in the standard MS medium supplemented with 15 μM BAP and 15 μM NAA (49.73 mg/g) (Figures 3 and S1). The lowest amount of phenolic compounds was found in control 1 and in the sample cultured on the medium with the NH_4^+/K^+ ratio of 0.33, BAP/NAA of 1.0, with the lowest concentration of plant growth regulators, i.e., 5 μM. The level of biosynthesis of TPCs on the nutrient medium with the highest increase in callus biomass was 13% lower than that on the standard MS medium (42.85 and 49.73 mg/g, respectively) (Figure 3).

The treatment of callus cultures with an elicitor led to qualitative and quantitative changes in the profile of phenolic compounds (Figures 3 and S2). MJ added at 100 μM gave a 119% increase relative to nonelicited cultures. We demonstrated that when callus cultures were exposed to an elicitor, four phenolic compounds emerged that were absent in nonelicited cultures; retention times were 13.6, 30.0, 33.3 and 46.0 (Table S1).

Multivariate statistical analysis was performed to classify the differences in phenolic compounds among callus cultures. Unsupervised principal component analysis (PCA) revealed two major clusters (Figure 4). Two principal components, PC1 and PC2, together accounted for 63% of the total variance. Control sample 1, sample No. 7 and samples treated with MJ can be clearly distinguished from the main cluster. Sample No. 7 is characterized by richer qualitative and quantitative composition (TPCs at 72.27 mg/g, 17 phenolic compounds).

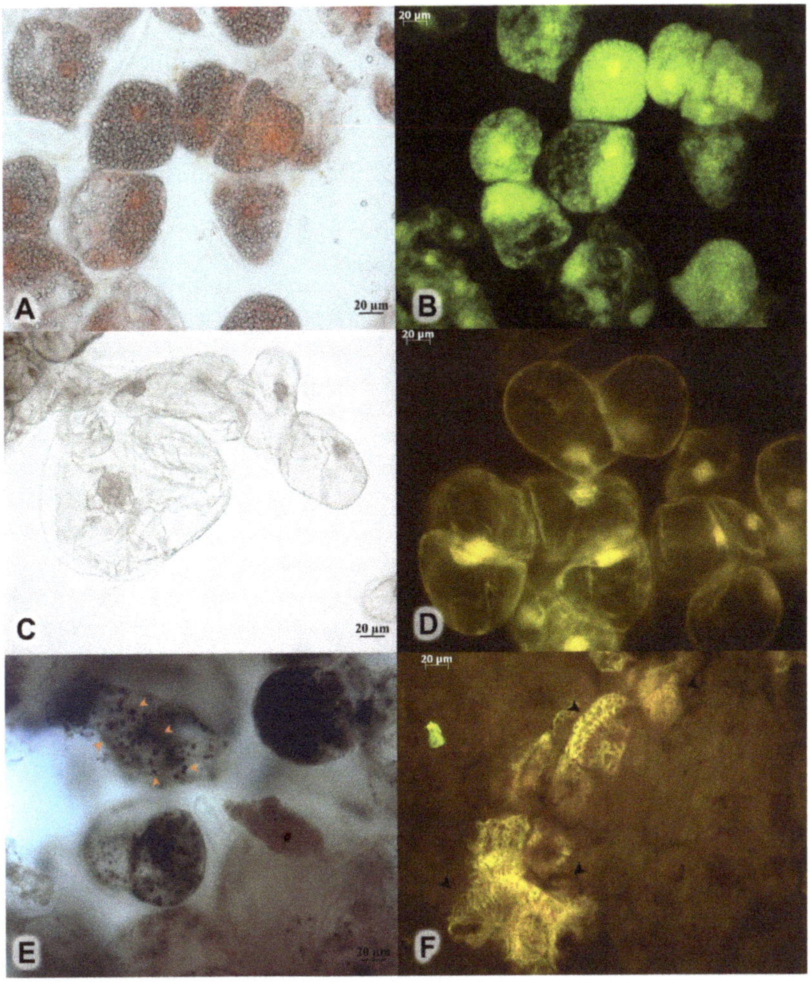

Figure 2. Callus culture of *R. rosea* cultivated on a modified MS medium at the NH_4^+/K^+ ratio of 0.33, 40 mM NO_3^-, the BAP/NAA ratio of 1.0, and concentrations of plant growth regulators at 30 µM. (**A**) Numerous small vacuoles in immature cells seen under a light microscope and (**B**) under a fluorescence microscope. (**C**) Mature cells with a single vacuole under the light microscope and (**D**) under the fluorescence microscope. (**E**) Lugol's solution for starch (orange arrows). (**F**) Fragments of spiral vessels as seen under the fluorescence microscope (black arrows).

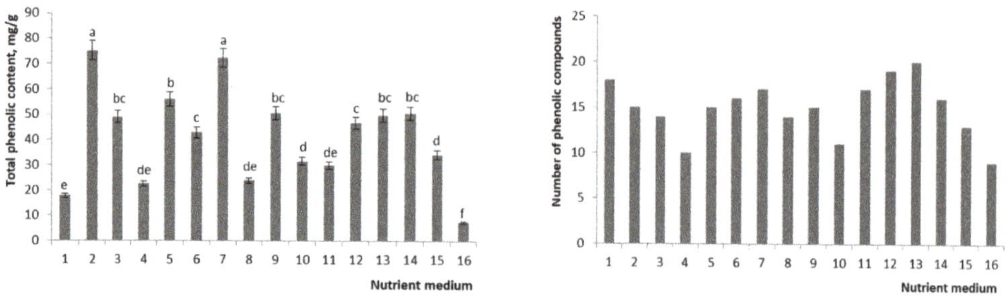

(A) (B)

Figure 3. Effect of composition of the nutrient medium and different concentrations of MJ on (**A**) total phenolic content and (**B**) number of phenolic compounds in callus cultures of *R. rosea*. Values are mean ± standard error (vertical error bars) of three replicates. Means with similar letters are not significantly different at $p \leq 0.05$ according to LSD test. Legend. Treatment group 1: NH_4^+/K^+ 0.33, NO_3^- 20 mM, BAP/NAA 1.0, BAP + NAA 5 μM; 2: NH_4^+/K^+ 0.33, NO_3^- 20 mM, BAP/NAA 0.33, BAP + NAA 30 μM; 3: NH_4^+/K^+ 0.33, NO_3^- 20 mM, BAP/NAA 1.0, BAP + NAA 30 μM; 4: NH_4^+/K^+ 0.33, NO_3^- 40 mM, BAP/NAA 0.33, BAP + NAA 17.5 μM; 5: NH_4^+/K^+ 0.33, NO_3^- 40 mM, BAP/NAA 0.33, BAP + NAA 30 μM; 6: NH_4^+/K^+ 0.33, NO_3^- 40 mM, BAP/NAA 1.0, BAP + NAA 30 μM; 7: NH_4^+/K^+ 0.33, NO_3^- 60 mM, BAP/NAA 0.33, BAP + NAA 30 μM; 8: NH_4^+/K^+ 0.33, NO_3^- 60 mM, BAP/NAA 1.0, BAP + NAA 30 μM; 9: NH_4^+/K^+ 1.0, NO_3^- 20 mM, BAP/NAA 0.33, BAP + NAA 17.5 μM; 10: NH_4^+/K^+ 1.0, NO_3^- 20 mM, BAP/NAA 0.33, BAP + NAA 30 μM; 11: NH_4^+/K^+ 1.0, NO_3^- 20 mM, BAP/NAA 1.0, BAP + NAA 30 μM; 12: NH_4^+/K^+ 1.0, NO_3^- 30 mM, BAP/NAA 1.0, BAP + NAA 17.5 μM; 13: NH_4^+/K^+ 1.0, NO_3^- 40 mM, BAP/NAA 1.0, BAP + NAA 30 μM; 14: NH_4^+/K^+ 0.33, NO_3^- 40 mM, BAP/NAA 1.0, BAP + NAA 30 μM, MJ 100 μM; 15: NH_4^+/K^+ 0.33, NO_3^- 40 mM, BAP/NAA 1.0, BAP + NAA 30 μM, MJ 200 μM; 16: control 1.

2.5.2. The Profile and Levels of Phenolic Compounds in Immature Plants (Control 2)

Five-year-old cultivated plants of *R. rosea* served as control 2 in the experiment when the biochemical composition was analyzed. HPLC (suitable for isolating a complex set of phenolic compounds) detected 10 phenolic compounds in the rhizomes and roots and 21 phenolic compounds in the above-ground part of *R. rosea* (Table 2). Compounds No. 5–8, identified as cinnamyl alcohol and its derivatives (phenylpropanoids), were found only in underground organs. Their observed maxima are typical for rhizomes: rosarin 16.33 mg/g, rosavin 41.73, rosin 25.10, and cinnamyl alcohol 42.74 mg/g (Table 2, Figure S3). The total content of phenolic compounds was 159.09 mg/g in the rhizomes and 88.68 mg/g in the roots.

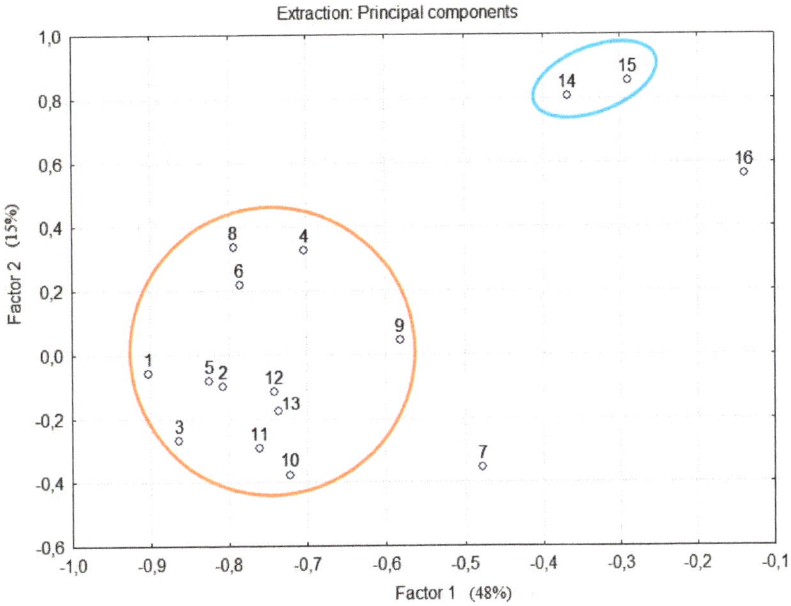

Figure 4. The principal component analysis (PCA) plot showing two clusters of callus cultures and control 1 of *R. rosea*. Blue ellipse: callus culture after treatment with MJ; orange ellipse: all other callus cultures in the experiment, except No. 7; No. 7: callus culture obtained on the medium that had NH_4^+/K^+ of 0.33, NO_3^- at 60 mM, BAP/NAA 0.33, BAP + NAA 30 μM; No. 16: control 1. Note: see Figure 3.

Table 2. Characteristics and levels of the phenolic compounds detected by HPLC in the extracts from rhizomes and roots of *R. rosea*.

Compound	Spectral Characteristics: λ_{max}, nm	Retention Time (t_R), min	Content, mg/g of Air-Dried Material Rhizomes	Roots
Gallic acid	272	1.8	26.15	21.20
Compound 2	216,280	2.9	3.20	2.30
Compound 3	228,296	6.5	0.51	0.23
Compound 4	218,274	8.5	0.41	-
Rosarin	253	10.7	16.33	7.41
Rosavin	253	12.5	41.73	11.73
Rosin	253	13.5	25.10	11.25
Cinnamyl alcohol	205,253	24.3	42.74	32.10
Rhodiosin	277,333,385	39.9	0.81	0.72
Rhodionin	277,333,385	40.5	2.11	1.74
	TPC		159.09	88.68

"-": not detected.

Our study revealed that none of the callus culture samples, including the MJ-treated one, control 1, and above-ground parts of *R. rosea* contained rosin, rosavin, rosarin, and cinnamyl alcohol.

3. Discussion

Using DoE, correct nutrient combinations were identified, along with the effects of their interactions with the other environmental parameters: plant growth regulators.

Similar methods have recently been used for optimizing callus cultures of various plant species [35,37,38]. This method has two advantages over the more common factorial design: DoE allows us to investigate the effects of independent variables between the actual experimental data points and allows a researcher to easily increase the number of experimental variables to more than five, which is not practical in typical factorial designs. Our findings may facilitate the application of DoE to tissue culture optimization and the in vitro production of secondary metabolites, where an understanding of the complex interactions among plant growth regulators and growth medium nutrients is sought. Nitrogen quantity and form have been the subject of many growth medium optimization studies [39–43]. The optimum nitrate concentration is reported to be 20–30 mM for both growth and taxol production in cell cultures of *Taxus yunnanensis* Cheng et al. [40]. In an experiment on the NH_4^+/NO_3^- ratio in the culture medium, ginsenoside production in the adventitious roots of *Panax ginseng* C.A. Meyer was affected by NH_4^+/NO_3^- ratios in the culture medium, showing the greatest productivity at 18.5 mM NO_3^- without NH_4^+ [39]. Biomass growth and azadirachtin production of *Azadirachta indica* (A. Juss) suspension cultures are significantly improved in a medium with a high NH_4^+/NO_3^- ratio [41]. According to our data, the NH_4^+/K^+ ratio is a crucial factor for biomass accumulation and production of TPCs in the callus culture of *R. rosea*. The optimal response was seen at the NH_4^+/K^+ ratio of 0.33 and 20–40 mM NO_3^-.

Plant growth regulators are one of the most important factors owing to their important regulating role in plant physiology and biochemistry [44]. An appropriate proportioning of cytokinins and auxins can maintain the balance between differentiation and dedifferentiation and achieve the goal of rapid proliferation for plant cells in vitro [45]. To obtain callus cultures of *Rhodiola* species, cytokinin BAP is most often employed in combination with various auxins. Compact callus aggregate suspension cultures of *Rhodiola imbricata* are obtained on the MS medium supplemented with 3 mg/L NAA and 3 mg/L BAP [28]. *Rhodiola quadrifida* (Pall.) calluses are obtained from hairy roots in the MS medium with the addition of hormones: 3 mg/L 2,4-D and 0.5 mg/L BAP [31]. Calluses of *R. sachalinensis* Boriss. can be successfully cultivated on the MS medium supplemented with 3 mg/L BAP and 0.3 mg/L NAA [27]. The medium containing 1 mg/L 2,4-D, 2 mg/L NAA, 0.5 mg/L BAP, and 0.1 mg/L kinetin proved to be the best for the induction of the callus from *R. quadrifida* (the induction rate was 83.3%); the optimized combination of plant growth regulators for callus subculture is 1 mg/L 2,4-D, 0.1 mg/L BAP, and 0.5 mg/L kinetin [23]. Several other combinations of plant growth regulators have also been found to be effective for callus induction in species of the genus *Rhodiola*. One research group revealed that callus induction of *R. imbricata* is frequently achieved in juvenile leaves (100% frequency) and roots (87.50%) in the MS medium supplemented with 0.5 mg/L thidiazuron and 1 mg/L NAA [30]. In another study, to obtain a callus culture of *R. rosea*, leaves were placed on the surface of a fresh MS medium supplemented with 3 mg/L of N^6-(2-isopentenyl) adenine and 0.3 mg/L IAA [29]. For callus subcultures and subsequent cell suspension cultures of *R. crenulata* L., full-strength MS containing 0.5 mg/L thidiazuron and 0.5 mg/L NAA turned out to be the best [26]. This paper optimized plant growth regulator proportioning, and maximum biomass and production of phenolic compounds were obtained in the medium containing BAP/NAA ratios of 0.33–1.00, provided that the concentration of plant growth regulators was 30 μM. Here, we found that optimal growth of a callus culture of *R. rosea* requires BAP > 0. Other reports suggest that the synergistic combinations of auxin and cytokinin can significantly alter the production of secondary metabolites depending on plant species [46]. When compared with the initial medium (control medium), total plant growth regulators content has not been changed. However, the selected complex of optimal factors, in general, contributed to a significant increase in the growth parameters of *R. rosea* calluses.

Here, the callus obtained under optimal culture conditions (NH_4^+/K^+ 0.33, BAP/NAA 1.0, BAP + NAA 30 μM, and NO_3^- 40 mM) was semi-friable yellowish green. The increase in the growth of fresh biomass on this medium was up to 703%, which is 2.7 times higher than the growth rates on the standard MS medium. Nonetheless, the level of TPC biosynthesis on the nutrient medium that gave the highest increase in callus biomass was 13% lower than that on the standard MS medium. The highest content of TPCs was observed in calluses grown on media with the NH_4^+/K^+ ratio of 0.33 and BAP/NAA of 0.33 and a concentration of plant growth regulators of 30 μM. Data on the profile and concentrations of phenolic compounds in in vitro cultures vary and sometimes are contradictory because such results are influenced by various factors and stages of plant development [47]. Using RSM, it has been reported that higher KH_2PO_4 depletion and 75 μM $m^{-2}s^{-1}$ light intensity favored the biosynthesis of anthocyanins and the other phenolic compounds and resulted in elevated antioxidant capacity in grape (Bogazkere Cv.) callus culture [48]. Through Plackett–Burman's design and RSM, optimal proportions of plant growth regulators for a cell suspension culture of *Siraitia grosvenorii* were obtained. With the optimized plant growth regulators, the obtained cell biomass and polyphenols content were 32.18% and 13.86%, respectively, more than plant growth regulators proportioning before optimization [49]. Using HPLC, we determined that the profile and levels of phenolic compounds were similar between the above-ground part of intact plants and the callus culture. TPC concentration varied among callus cultures from 14.9 to 71.6 mg/g, the number of phenolic compounds from 10 to 20; in the above-ground part, these values were 73.1 mg/g and 21 phenolic compounds, respectively. Several new phenolic compounds were identified in the callus cultures: compounds No. 9, 11, 19, 26, and 27 (see Table S1).

Jasmonic acid is thought to be involved in the signal transduction pathway that induces the production of defense compounds in plants, such as alkaloids, terpenoids, and polyphenols [50]. MJ is an effective elicitor that participates in plant defense response pathways and triggers plant metabolite biosynthesis. Accordingly, MJ has been used for inducing metabolite production in plant cell cultures [33]. In the present study, the medium was supplemented with different concentrations of MJ in 45-day-old cultures, and phenolic-compound accumulation was determined after three days of cultivation. The treatment of callus cultures with an elicitor led to qualitative and quantitative changes in the profile of phenolic compounds. In this case, MJ concentration was of paramount importance. The use of 100 μM MJ was optimal and led to an increase in the TPC content up to 47.9 mg/g. We showed that when callus cultures are exposed to an elicitor, phenolic compounds emerge that are absent in the sample without treatment; their retention times are 13.6, 30.0, 33.3, and 46.0 min. The effectiveness of MJ as an elicitor has been demonstrated for many in vitro cultures, including those of *Rhodiola* species [28,51–57]. For example, for *Rhodiola* it is reported that the levels of bioactive compounds increase with MJ supplementation in a dose-dependent manner. The highest salidroside content (4.75 mg/g dry weight) is obtained during treatment with MJ at 125 μM [28].

According to our HPLC analyses of rhizomes, the levels of rosarin, rosavin, rosin, and cinnamyl alcohol are 16.33, 41.73, 25.10, and 42.74 mg/g, respectively, while in the roots, their respective levels are 7.41, 11.73, 11.25 and 32.10 mg/g. In the aerial parts (leaves and stems), seedlings, and callus cultures, no detectable rosavins were found, in line with a report of Peschel et al. [58], wherein the aerial parts of wild *R. rosea*, no rosavins were identified, while the content of salidroside was below the detection limit. Although salidroside (0.53%) was found in the leaves of *R. rosea* from Rila Mountain, Bulgaria [24] and in the leaves and stems of *R. rosea* cultivated in Poland (salidroside 0.12% and total rosavins 0.3%), the aerial parts of the plant grow anew every year and therefore their content is consistent each time [59]. Rattan et al. [30] found that rosavin and rosarin are present at the highest concentration in root-derived compact green calluses (0.15 mg/g dry weight) and root-derived friable green calluses (0.07 mg/g dry weight). Kurkin et al. [60] noted that in a suspension culture of *R. rosea*, the main phenolic compound is a phenylpropanoid called triandrin, while in a callus culture, the process of biosynthesis went further and, together with triandrin, the major phenolic compounds were dimeric phenylpropanoids: lignans; in other words, "ageing" of the biomass took place. Those authors emphasized the finding that neither salidroside nor phenylpropanoids—which are characteristic for the rhizomes of roseroot stonecrop (rosin, rosavin, and rosarin)—were found in the samples of biomass. In the in vitro cultures, a significant enhancement in the production of rosin and its derivatives was observed when the cultures were fed with the precursor: cinnamyl alcohol [5]. In *R. rosea* compact callus aggregate cultures, the observed rosin and rosarin content was even higher than that in field-cultivated plants, while the rosavin level was five times lower [2]. Our next research project on the callus culture of *R. rosea* will be aimed at finding an optimal concentration and duration of cinnamyl alcohol treatment to promote the biosynthesis of phenylpropanoids.

4. Materials and Methods

4.1. Plant Material

R. rosea seeds were collected from its natural habitat on the southern slopes of the Iolgo ridge, Karakol lakes, Altai Republic (Russia). Altitude was 1800–2000 m a.s.l. *R. rosea* samples growing in this area belong to the ecotype of moderately humid habitats with moderate soil moisture (44–50%). This ecotype was previously chosen by us as optimal for the introduction experiment in the conditions of the forest-steppe zone of Western Siberia (Russia). Our earlier studies confirm the high biological and biosynthetic potential of these samples [61]. In vitro cultivation and ex vitro acclimatization of *R. rosea* plants were performed in the Laboratory of Biotechnology, CSBG SB RAS (Novosibirsk, Russia) according to the previously developed method [62].

4.2. Establishment of Callus Cultures

A callus line was developed from cotyledon explants of in vitro-grown seedlings of *R. rosea*. Seeds were germinated in the MS [63] basal medium without plant growth regulators but supplemented with 30 g/L sucrose and 6 g/L agar. For callus induction, cotyledon explants were excised from 21-day-old seedlings and placed onto the MS medium supplemented with 15 µM BAP (Sigma-Aldrich, St Louis, MO, USA), 15 µM NAA (Sigma-Aldrich), 30 g/L sucrose, and 6 g/L agar. The conditions of explant cultivation were as follows: photoperiod, 16/8-h light/dark cycle; illumination intensity, 2–3 klx; and temperature, 24 ± 1 °C. Callus cultures were maintained on the MS medium containing 15 µM BAP and 15 µM NAA via regular subculturing with a 45-day interval.

For the establishment of callus cultures at different media compositions, an actively growing 6-month-old callus was used as a source. The experiment lasted 45 days. Fresh and dry weights were measured by taking the average of three plates for each treatment type. Percent increase in fresh weight was calculated from the initial weight of the subcultured callus.

4.3. Optimal (Combined) Design

The experiment had a mixture–concentration design and included four mixture components (K^+, NH_4^+, BAP, and NAA) and two numerical factors (NO_3^- and BAP + NAA concentrations). As NH_4^+/K^+ and BAP/NAA were regarded as components of mixture1 and mixture2, the range for each component was expressed as a proportion; all component proportions in each mixture add up to 1.0. NO_3^- concentration ranged from 20 to 60 mM, K^+ proportion from 0.5 to 1.0, and NH_4^+ proportion ranged from 0 to 0.5. BAP + NAA concentration ranged from 5 to 30 µM, NAA proportion from 0.5 to 1.0, and BAP proportion from 0 to 0.5. The concentration of K^+ plus NH_4^+ matched the NO_3^- concentration to maintain charge neutrality. Design points were selected via D-optimal criteria to satisfy a quadratic polynomial for the mixtures (NH_4^+/K^+ and BAP/NAA) and the numerical factors (NO_3^- and BAP + NAA) in various combinations in the mixture. The experiment included a data point for the MS basal medium.

All solution recipes were derived using the linear programming approach described by Niedz and Evens [64]. The concentrations of salts/acids/bases needed for each data point in the design space were calculated in the ARS-Media (Version 1.0) software (ion solution calculation program), which is available as a free download via http://www.ars.

usda.gov/services/software/download.htm?softwareid=148 (accessed on 10 September 2020). For each treatment, all ions present and their concentrations were entered into ARS-Media. Ions other than those being varied were fixed at their normal levels present in the MS medium.

4.4. *Effect of MJ Treatment*

Callus cultures with the highest biomass growth were used for elicitation. Subsequently, a 45-day-old callus was transferred to a 100 mL Erlenmeyer flask containing 20 mL of a modified MS medium (NH_4^+/K^+ 0.33, NO_3^- 40 mM) supplemented with 15 µM NAA, 15 µM BAP, and different concentrations of MJ (100 or 200 µM). The modified MS medium without MJ served as a control. The culture flasks were placed on a rotary shaker (90 rpm, Elmi, S-3-02L, Latvia) at 24 ± 2 °C under a 16 h/8 h photoperiod. The accumulation of biomass and the production of phenolic compounds were implemented after 3 days of cultivation.

4.5. *Biochemical Analysis*

4.5.1. Extraction

The dry raw material (callus cultures, in vitro seedlings, rhizomes, roots, and aboveground parts) was crushed to a particle size of 1 mm, mixed, and average samples were collected. Biochemical analysis was performed only on well-growing callus cultures with an increase in wet biomass of more than 300% and for callus grown on the standard MS medium. Double extraction was performed to isolate phenolic compounds. A 0.5 g sample of the crushed material was extracted with 30 mL of aqueous 50% ethanol for 8 h, and then with 20 mL of 70% ethanol for 50 min in a water bath. Each filter cake was washed with 5 mL of 70% ethanol. The combined extract was concentrated to 20 mL. To remove impurities, 1 mL of the extract was diluted with double-distilled water to 5 mL and passed through a Diapak C16 (ZAO BioKhimMak, Moscow, Russia) concentrating cartridge.

4.5.2. HPLC Analysis

The profile of phenolic compounds in the samples was investigated by HPLC using an Agilent 1200 liquid chromatograph equipped with a diode array detector, Zorbax SB–C18 column (4.6 × 150 mm, with sorbent grain size 5 µm; Agilent Technologies, Santa Clara, CA, USA). In the mobile phase, the methanol content in the aqueous solution of phosphoric acid (0.1%) was varied from 50% to 52% within 56 min. The flow rate of the eluent was 1 mL/min; the column temperature was 26 °C and the volume of the injected sample was 10 µL. Detection was carried out at λ = 220, 250, 270, 290, 360, and 370 nm. Concentrations of substances were calculated by detection at 255 nm. Methyl alcohol (extra pure grade), orthophosphoric acid (extra pure grade), and double-distilled water were utilized to prepare mobile phases. Standards from Aobious (USA) and Sigma-Aldrich (Germany) were used for identification. Standard solutions were prepared at a concentration of 10 µg/mL in ethyl alcohol. The retention time of the peaks of compounds in the chromatograms of the analyzed samples and their UV spectra were compared with those of the standard samples. Quantitative analysis of individual phenolic compounds in plant samples was performed by the external standard method (% of the weight of air-dry raw materials).

To assess biosynthetic potential of the callus cultures, two controls were used in the study. Control 1 was 21-day-old in vitro seedlings of *R. rosea*. Control 2 consisted of cultivated 5-year-old *R. rosea* plants propagated in an in vitro culture [36]. The initial material for the control samples was seeds from the same population and the same year of collection as the experimental samples.

4.6. Histochemical Analysis

Callus cultures of *R. rosea* were fixed in a mixture of formalin, acetic acid, and 70% ethanol (7:7:100 $v/v/v$) for 4 days and then stored in 70% (v/v) ethanol. For fluorescent microscopy, fresh calluses were used.

For histochemical characterization, the calluses of *R. rosea* fixed as described above were sectioned (25–35 µm) on a MICROM HM 430 microtome (Thermo Fisher Scientific, Munich, Germany) with fast freezing unit KS 34 S (Thermo Fisher Scientific, Munich, Germany). For light microscopy, sections were analyzed by means of a Carl Zeiss Axioscope A1 microscope equipped with digital camera Axiocam 506 color, and the ZEN 2012 software (blue edition) (Carl Zeiss Ltd., Herts, UK) and Carl Zeiss Primo Star iLED equipped with a filter system (470 nm), digital camera AxioCam MRc, and the AxioVision 4.8 software.

Cross-sections were investigated by the following histochemical tests: a ferric chloride reaction to detect phenolic compounds and a reaction with Lugol's solution to detect starch. For fluorescence microscopy, fresh sections were treated using a 2% (w/v) solution of safranin for 2 min to detect starch, inducing yellow color at 470 nm, respectively. For autofluorescence examination, the sections were directly viewed under LED light at 470 nm.

4.7. Statistical Data Analysis

Design-Expert® 13 was used for experimental design construction, model evaluation, and all analyses. The data expressed as mean and standard error (M $\pm$ SE) were subjected to ANOVA in the STATISTICA 6.0 software. The differences between means were tested for significance by the LSD test at $p \leq 0.05$. In addition, clustering was performed by PCA.

5. Conclusions

In summary, this is the first report describing a statistical optimization study of biomass accumulation and the production of phenolic compounds in callus cultures of *R. rosea* using DOE. Computer-generated optimal design is an excellent tool for reducing treatment numbers compared to traditional factorial designs. The overall effect of the resulting model was highly significant ($p < 0.0001$), indicating that studied factors (NH_4^+, K^+, NO_3^-, BAP, and NAA) significantly affected the growth of callus cultures of *R. rosea*. The best callus growth (703%) and the highest production of TPCs (75.17 mg/g) were achieved at an NH_4^+/ K^+ ratio of 0.33 and BAP/NAA of 0.33, provided that the concentration of plant growth regulators was 30 µM and that of NO_3^- was $\leq$40 mM. The information presented in this study may be useful for future research related to the cell culture of *R. rosea* and other *Rhodiola* species.

Supplementary Materials: The following supporting information can be downloaded at: https://www.mdpi.com/article/10.3390/plants11010124/s1, Figure S1: A chromatogram of a water-ethanol extract of the *R. rosea* callus culture that featured the highest TPC concentration, Figure S2: A chromatogram of a water-ethanol extract of the *R. rosea* callus culture cultivated under optimal nutrient conditions (A) with MJ 100 µM and (B) without MJ, Figure S3: A chromatogram of a water-ethanol extract of *R. rosea* rhizomes; Table S1: Characteristics and levels of the phenolic compounds detected by HPLC in the extracts from callus cultures and above-ground parts (herb) of *R. rosea* (mg/g).

Author Contributions: Conceptualization, A.A.E. and E.V.B.; methodology, A.A.E., A.A.P. and N.V.F.; validation, D.A.K. and A.S.E.; formal analysis, S.Y.M.; investigation, A.A.E. and A.A.P.; writing—original draft preparation, A.A.E. and A.A.P.; writing—review and editing, E.V.B. and M.S.K.; visualization, A.S.E.; project administration, E.V.B. All authors have read and agreed to the published version of the manuscript.

Funding: This research was funded by the Russian Foundation for Basic Research and Novosibirsk region, grant number 20-44-540005, the budgetary project of the Central Siberian Botanical Garden, SB RAS No. AAAA-A17-117012610051-5 within the framework of the State Assignment and the Ministry of Science and Higher Education of the Russian Federation, theme No. 121041200194-7 (IPP RAS).

Institutional Review Board Statement: Not applicable.

Informed Consent Statement: Not applicable.

Data Availability Statement: Data available on request due to restriction.

Acknowledgments: To prepare the publication, materials of the bioresource scientific collection of the Central Siberian Botanical Garden of the SB RAS "Collection of living plants indoors and outdoors", USU_440534 were used.

Conflicts of Interest: The authors declare no conflict of interest.

References

1. Marchev, A.; Haas, C.; Schulz, S. Sage in vitro cultures: A promising tool for the production of bioactive terpenes and phenolic substances. *Biotechnol. Lett.* **2014**, *36*, 211–221. [CrossRef]
2. Grech-Baran, M.; Sykłowska-Baranek, K.; Pietrosiuk, A. Biotechnological approaches to enhance salidroside, rosin and its derivatives production in selected *Rhodiola* spp. in vitro cultures. *Phytochem. Rev.* **2015**, *14*, 657–674. [CrossRef]
3. Zheleznichenko, T.; Banaev, E.; Asbaganov, S.; Voronkova, M.; Kukushkina, T.; Filippova, E.; Mazurkova, N.; Shishkina, L.; Novikova, T. *Nitraria schoberi* L. hairy root culture as a source of compounds with antiviral activity against influenza virus subtypes A(H5N1) and A(H3N2). *3 Biotech* **2018**, *8*, 260. [CrossRef] [PubMed]
4. Chiang, H.M.; Chen, H.C.; Wu, C.S.; Wu, P.Y.; Wen, K.C. *Rhodiola* plants: Chemistry and biological activity. *J. Food Drug Anal.* **2015**, *23*, 359–369. [CrossRef]
5. Marchev, A.S.; Dinkova-Kostova, A.T.; György, Z. *Rhodiola rosea* L.: From golden root to green cell factories. *Phytochem. Rev.* **2016**, *15*, 515–536. [CrossRef]
6. Khanna, K.; Mishra, K.P.; Ganju, L.; Singh, S.B. Golden root: A wholesome treat of immunity. *Biomed. Pharmacother* **2017**, *87*, 496–502. [CrossRef] [PubMed]
7. Brinckmann, J.A.; Cunningham, A.B.; Harter, D.E.V. Running out of time to smell the roseroots: Reviewing threats and trade in wild *Rhodiola rosea* L. *J. Ethnopharmacol.* **2021**, *269*, 113710. [CrossRef] [PubMed]
8. György, Z.; Tolonen, A.; Pakonen, M.; Neubauer, P.; Hohtola, A. Enhancement of the production of cinnamyl glycosides in CCA cultures of *Rhodiola rosea* through biotransformation of cinnamyl alcohol. *Plant Sci.* **2004**, *166*, 229–236. [CrossRef]
9. Ma, L.Q.; Gao, D.Y.; Wang, Y.N.; Wang, H.H.; Zhang, J.X.; Pang, X.B.; Hu, T.S.; Lu, S.Y.; Li, G.F.; Ye, H.C.; et al. Effects of overexpression of endogenous phenylalanine ammonia-lyase (PALrs1) on accumulation of salidroside in *Rhodiola sachalinensis*. *Plant Biol.* **2008**, *10*, 323–333. [CrossRef]
10. György, Z.; Hohtola, A. Production of cinnamyl glycosides in compact callus aggregate cultures of *Rhodiola rosea* through biotransformation of cinnamyl alcohol. In *Protocols for In Vitro Cultures and Secondary Metabolite Analysis of Aromatic and Medicinal Plants. Methods in Molecular Biology*; Jain, S.M., Saxena, P., Eds.; Humana Press: New York, NY, USA, 2009; Volume 547, pp. 305–312.
11. Martin, J.; Pomahacova, B.; Dusek, J.; Duskova, J. In vitro culture establishment of *Schizandra chinensis* (Turz.) Baill. and *Rhodiola rosea* L., two adaptogenic compounds producing plants. *J. Phytol.* **2010**, *2*, 80–87.
12. Furmanowa, M.; Hartwich, M.; Alfermann, A.W. Rosavin as a product of glycosylation by *Rhodiola rosea* (roseroot) cell cultures. *Plant Cell Tissue Organ Cult.* **1999**, *56*, 105–110. [CrossRef]
13. Grech-Baran, M.; Sykłowska-Baranek, K.; Krajewska-Patan, A. Biotransformation of cinnamyl alcohol to rosavins by non-transformed wild type and hairy root cultures of *Rhodiola kirilowii*. *Biotechnol. Lett.* **2014**, *36*, 649–656. [CrossRef]
14. Kapoor, S.; Raghuvanshi, R.; Bhardwaj, P.; Sood, H.; Saxena, S.; Chaurasia, O.P. Influence of light quality on growth, secondary metabolites production and antioxidant activity in callus culture of *Rhodiola imbricata* Edgew. *J. Photochem. Photobiol. B* **2018**, *183*, 258–265. [CrossRef] [PubMed]
15. Condra, L.W. *Reliability Improvement with Design of Experiments*; Marcel Dekker AG: New York, NY, USA, 2001.
16. Niedz, R.P.; Evens, T.J. Design of experiments (DoE)—History, concepts, and relevance to in vitro culture. *In Vitro Cell. Dev. Biol. Plant* **2016**, *52*, 547–562. [CrossRef]
17. Niedz, R.P.; Evens, T.J. The effects of nitrogen and potassium nutrition on the growth of nonembryogenic and embryogenic tissue of sweet orange (*Citrus sinensis* (L.) Osbeck). *BMC Plant Biol.* **2008**, *8*, 126. [CrossRef] [PubMed]
18. Butenko, R.G. *The Culture of Isolated Tissues and the Physiology of Plant Morphogenesis*; Nauka Publications: Moscow, Russia, 1964; 350p.
19. Ramage, C.M.; Williams, R.R. Inorganic nitrogen requirements during shoot organogenesis in tobacco leaf discs. *J. Exp. Bot.* **2002**, *53*, 1437–1443. [CrossRef] [PubMed]
20. Cousson, A.; Van Tran Thanh, K. Influence of ionic composition of the culture medium on de novo flower formation in tobacco thin cell layers. *Can. J. Bot.* **1993**, *71*, 506–511. [CrossRef]
21. Poothong, S.; Reed, B.M. Optimizing shoot culture media for Rubus germplasm: The effects of NH_4^+, NO_3^-, and total nitrogen. *In Vitro Cell. Dev. Biol. Plant* **2016**, *52*, 265–275. [CrossRef]
22. Niedz, R.P.; Evens, T.J. Regulating plant tissue growth by mineral nutrition. *In Vitro Cell. Dev. Biol. Plant* **2007**, *43*, 370–381. [CrossRef]

23. Sheng, C.Z.; Hu, T.Q.; Bi, H.; Yuan, Y.J.; Jiang, Y. Effects of plant growth substances on induction and culture of callus from *Rhodiola quadrifida*. *Zhongguo Zhong Yao Za Zhi* **2005**, *30*, 1237–1240.
24. Tasheva, K.; Kosturkova, G. Bulgarian golden root in vitro cultures for micropropagation and reintroduction. *Cent. Eur. J. Biol.* **2010**, *5*, 853–863. [CrossRef]
25. Ghiorghită, G.; Hârtan, M.; Maftei, D. Some considerations regarding the in vitro culture of *Rhodiola rosea* L. *Rom. Biotechnol. Lett.* **2011**, *16*, 5902–5908.
26. Shi, L.; Wang, C.; Zhou, X.; Zhang, Y.; Liu, Y.; Ma, C. Production of salidroside and tyrosol in cell suspension cultures of *Rhodiola crenulata*. *Plant Cell Tiss. Organ. Cult.* **2013**, *114*, 295–303. [CrossRef]
27. Li, Y.; Shao, C.H.; Park, S.Y. Production of salidroside and polysaccharides in *Rhodiola sachalinensis* using airlift bioreactor systems. *Acta Physiol. Plant* **2014**, *36*, 2975–2983. [CrossRef]
28. Kapoor, S.; Sharma, A.; Bhardwaj, P. Enhanced production of phenolic compounds in compact allus Aggregate suspension cultures of *Rhodiola imbricata* Edgew. *Appl. Biochem. Biotechnol.* **2019**, *187*, 817–837. [CrossRef]
29. Mirmazloum, I.; Kiss, A.; Ladányi, M. Production of cinnamyl alcohol glycosides by biotransformation in roseroot callus cells. *Plant Cell Tiss. Organ Cult.* **2019**, *139*, 29–37. [CrossRef]
30. Rattan, S.; Rattan, S.; Sood, A.; Kumar, P.; Kumar, A.; Kumar, D.; Kumar, D.; Warghat, A.R.; Warghat, A.R. Phenylethanoids, phenylpropanoids, and phenolic acids quantification vis-à-vis gene expression profiling in leaf and root derived callus lines of *Rhodiola imbricata* (Edgew.). *Ind. Crops Prod.* **2020**, *154*, 112708. [CrossRef]
31. Stepanova, A.; Malunova, M.; Salamaikina, S.; Selimov, R.; Solov'eva, A. Establishment of *Rhodiola quadrifida* hairy roots and callus culture to produce bioactive compounds. *Phyton-Int. J. Exp. Bot.* **2021**, *90*, 543–552. [CrossRef]
32. Kolewe, M.E.; Gaurav, V.; Roberts, S.C. Pharmaceutically active natural product synthesis and supply via plant cell culture technology. *Mol. Pharm.* **2008**, *5*, 243–256. [CrossRef]
33. Zhao, J.; Davis, L.C.; Verpoorte, R. Elicitor signal transduction leading to production of plant secondary metabolites. *Biotechnol. Adv.* **2005**, *23*, 283–333. [CrossRef]
34. Gundlach, H.; Müller, M.J.; Kutchan, T.M.; Zenk, M.H. Jasmonic acid is a signal transducer in elicitor-induced plant cell cultures. *Proc. Natl. Acad. Sci. USA* **1992**, *89*, 2389–2393. [CrossRef]
35. Chu, U.C.; Adelberg, J.; Lowe, K. Use of DoE methodology to optimize the regeneration of high-quality, single-copy transgenic *Zea mays* L. (maize) plants. *In Vitro Cell. Dev. Biol. Plant* **2019**, *55*, 678–694. [CrossRef]
36. Erst, A.A.; Petruk, A.A.; Zibareva, L.N.; Erst, A.S. Morphological, histochemical and biochemical features of cultivated *Rhodiola rosea* (Altai Mountains ecotype). *Contemp. Probl. Ecol.* **2021**, *14*, 701–710. [CrossRef] [PubMed]
37. Gómez-Montes, E.O.; Oliver-Salvador, C.; Durán-Figueroa, N. Optimization of direct shoot regeneration using cotyledonary explants and true leaves from lettuce cv. Romaine (*Lactuca sativa* L.) by surface response methodology. *Plant Growth Regul.* **2015**, *77*, 327–334. [CrossRef]
38. Bagherieh-Najjar, M.B.; Nezamdoost, T. Optimization of shikonin production in *Onosma dichroantha* callus using response surface methodology. *Plant Cell Tiss. Organ Cult.* **2016**, *126*, 399–409. [CrossRef]
39. Yu, K.W.; Hahn, E.J.; Paek, K.Y. Effects of NH_4^+: NO_3^- ratio and ionic strength on adventitious root growth and ginsenoside production in bioreactor culture of *Panax ginseng* C.A. Meyer. *Acta Hortic.* **2001**, *560*, 259–262. [CrossRef]
40. Chen, Y.Q.; Yi, F.; Cai, M. Effects of amino acids, nitrate, and ammonium on the growth and taxol production in cell cultures of *Taxus yunnanensis*. *Plant Growth Regul.* **2003**, *41*, 265–268. [CrossRef]
41. Prakash, G.; Srivastava, A.K. Statistical media optimization for cell growth and azadirachtin production in *Azadirachta indica* (A. Juss) suspension cultures. *Process Biochem.* **2005**, *40*, 3795–3800. [CrossRef]
42. Irshad, M.; Debnath, B.; Mitra, S. Accumulation of anthocyanin in callus cultures of red-pod okra (*Abelmoschus esculentus* (L.) Hongjiao) in response to light and nitrogen levels. *Plant Cell Tiss. Organ Cult.* **2018**, *134*, 29–39. [CrossRef]
43. Oleszkiewicz, T.; Kruczek, M.; Baranski, R. Repression of carotenoid accumulation by nitrogen and NH_4^+ supply in carrot callus cells in vitro. *Plants* **2021**, *10*, 1813. [CrossRef]
44. Martinez, M.E.; Jorquera, L.; Poirrier, P.; Díaz, K.; Chamy, R. Effect of the carbon source and plant growth regulators (PGRs) in the induction and maintenance of an in vitro callus culture of *Taraxacum officinale* (L) weber Ex F.H. Wigg. *Agronomy* **2021**, *11*, 1181. [CrossRef]
45. Small, C.C.; Degenhardt, D. Plant growth regulators for enhancing revegetation success in reclamation: A review. *Ecol. Eng.* **2018**, *118*, 43–51. [CrossRef]
46. Goyal, S.; Ramawat, K.G. Synergistic effect of morphactin on cytokinin-induced production of isoflavonoids in cell cultures of *Pueraria tuberosa* (Roxb. ex. Willd.) DC. *Plant Growth Regul.* **2008**, *55*, 175–181. [CrossRef]
47. Muraseva, D.S.; Kostikova, V.A. In vitro propagation of *Spiraea betulifolia* subsp. *aemiliana* (Rosaceae) and comparative analysis of phenolic compounds of microclones and intact plants. *Plant Cell Tiss. Organ Cult.* **2021**, *144*, 493–504. [CrossRef]
48. Karaaslan, M.; Ozden, M.; Vardin, H.; Yilmaz, F.M. Optimisation of phenolic compound biosynthesis in grape (Bogazkere Cv.) callus culture. *Afr. J. Biotechnol.* **2013**, *12*, 3922–3933. [CrossRef]
49. Liu, Z.; Zhu, X.; Mohsin, A.; Yin, Z.; Zhuang, Y.; Zhou, B.; Du, L.; Yin, X.; Liu, N.; Wang, Z.; et al. Embryogenic callus induction, cell suspension culture, and spectrum-effect relationship between antioxidant activity and polyphenols composition of Siraitia grosvenorii cultured cells. *Ind. Crops Prod.* **2022**, *176*, 114380. [CrossRef]

50. Mizukami, H.; Tabira, Y.; Ellis, B.E. Methyl jasmonate-induced rosmarinic acid biosynthesis in *Lithospermum erythrorhizon* cell suspension cultures. *Plant Cell Rep.* **1993**, *12*, 706–709. [CrossRef] [PubMed]
51. Shabania, L.; Ehsanpoura, A.A.; Asgharib, G.; Emamib, J. Glycyrrhizin production by in vitro cultured *Glycyrrhiza glabra* elicited by methyl jasmonate and salicylic acid. *Russ. J. Plant Physiol.* **2009**, *56*, 621–626. [CrossRef]
52. Martin, K.P.; Sabovljevic, A.; Madassery, J. High-frequency transgenic plant regeneration and plumbagin production through methyl jasmonate elicitation from hairy roots of *Plumbago indica* L. *J. Crop. Sci. Biotech.* **2011**, *14*, 205–212. [CrossRef]
53. Ram, M.; Prasad, K.V.; Singh, S.K.; Hada, B.S.; Kumar, S. Influence of salicylic acid and methyl jasmonate elicitation on anthocyanin production in callus cultures of *Rosa hybrida* L. *Plant Cell Tiss. Organ Cult.* **2013**, *113*, 459–467. [CrossRef]
54. Chodisetti, B.; Rao, K.; Gandi, S.; Giri, A. Gymnemic acid enhancement in the suspension cultures of *Gymnema sylvestre* by using the signaling molecules–methyl jasmonate and salicylic acid. *In Vitro Cell. Dev. Biol. Plant* **2015**, *51*, 88–92. [CrossRef]
55. Erst, A.A.; Zibareva, L.N.; Filonenko, E.S.; Zheleznichenko, T.V. Influence methyl jasmonate on production of ecdysteroids from hairy roots of *Silene linicola* C.C.Gmelin. *Russ. J. Bioorg. Chem.* **2019**, *7*, 76–82. [CrossRef]
56. Ho, T.T.; Murthy, H.N.; Park, S.Y. Methyl jasmonate induced oxidative stress and accumulation of secondary metabolites in plant cell and organ cultures. *Int. J. Mol. Sci.* **2020**, *21*, 716. [CrossRef] [PubMed]
57. Erst, A.A.; Zibareva, L.N.; Filonenko, E.S. Variation in phytoecdysteroid accumulation in hairy roots of *Silene linicola* over extended time periods. *J. Plant Biochem. Biotechnol.* **2021**. [CrossRef]
58. Peschel, W.; Kump, A.; Horváth, A.; Csupor, D. Age and harvest season affect the phenylpropenoid content in cultivated European *Rhodiola rosea* L. *Ind. Crops Prod.* **2016**, *83*, 787–802. [CrossRef]
59. Kołodziej, B.; Sugier, D. Influence of plant age on the chemical composition of roseroot (*Rhodiola rosea* L.). *Acta Sci. Pol. Hortorum. Cultus* **2013**, *12*, 147–160.
60. Kurkin, V.A.; Zapesochnaya, G.G.; Dubichev, A.G. Phenylpropanoids of a callus culture of *Rhodiola rosea*. *Chem. Nat. Compd.* **1991**, *27*, 419–425. [CrossRef]
61. Erst, A.A.; Yakubov, V.V. Regenerative in vitro capacity of rare species *Rhodiola rosea* L. from various habitats. *Contemp. Probl. Ecol.* **2019**, *12*, 368–376. [CrossRef]
62. Erst, A.; Erst, A.; Shmakov, A. In vitro propagation of rare species *Rhodiola rosea* from Altai Mountains. *Turczaninowia* **2018**, *21*, 78–86. [CrossRef]
63. Murashige, T.; Skoog, F. A revised medium for rapid growth and bioassays with tobacco tissue cultures. *Physiol. Plant* **1962**, *15*, 473–497. [CrossRef]
64. Niedz, R.P.; Evens, T.J. A solution to the problem of ion confounding in experimental biology. *Nat. Methods* **2006**, *3*, 417. [CrossRef] [PubMed]

plants

MDPI

Article

Halophyte *Artemisia caerulescens* L.: Metabolites from In Vitro Shoots and Wild Plants

Ylenia Pieracci [1], Martina Vento [1], Luisa Pistelli [1,2], Tiziana Lombardi [3,4] and Laura Pistelli [2,3,4,*]

1 Dipartimento di Farmacia, Università di Pisa, Via Bonanno 6, 56126 Pisa, Italy; ylenia.pieracci@phd.unipi.it (Y.P.); m.vento1@studenti.unipi.it (M.V.); luisa.pistelli@unipi.it (L.P.)

2 Centro Interdipartimentale di Ricerca Nutraceutica e Alimentazione per la Salute 'NUTRAFOOD', Università di Pisa, Via del Borghetto 80, 56124 Pisa, Italy

3 Dipartimento di Scienze Agrarie, Alimentari e Agro-Alimentari, Università di Pisa, Via del Borghetto 80, 56124 Pisa, Italy; tiziana.lombardi@unipi.it

4 Centre for Climate Change Impact (CIRSEC), University of Pisa, Via del Borghetto 80, 56124 Pisa, Italy

* Correspondence: laura.pistelli@unipi.it

Abstract: Halophyte plants are potential resources to deal with the increasing soil salinity determined by climatic change. In this context, the present study aimed to investigate the germplasm conservation of *Artemisia caerulescens* collected in the San Rossore Estate (Pisa, Italy) through in vitro culture, biochemical properties, and the phytochemical composition of the volatile fraction of both in vitro shoots and different organs of wild plants (leaves, young and ripe inflorescences). The best medium tested for the shoot proliferation was MS, with the addition of 1 μM BA. Total polyphenol content and antioxidant activity were noticeable in both the inflorescences, while leaves and in vitro shoots showed lower amounts. Concerning the phytochemical investigation, the headspaces (HSs) and the essential oils (EOs) were characterized by oxygenated monoterpenes as the main chemical class of compounds in all samples, and with α- and β-thujone as the major constituents. However, the EOs were characterized by noticeable percentages of phenylpropanoids (23.6–28.8%), with brevifolin as the unique compound, which was not detected in the spontaneous volatile emissions of the same parts of the wild plant. Good amounts of EOs were obtained from different organs of the wild plant, comprising between 0.17% and 0.41% of the young and ripe inflorescences, respectively.

Keywords: shoot proliferation; polyphenols; antioxidant activity; essential oils; HS-SPME; GC-MS; PCA; HCA

Citation: Pieracci, Y.; Vento, M.; Pistelli, L.; Lombardi, T.; Pistelli, L. Halophyte *Artemisia caerulescens* L.: Metabolites from In Vitro Shoots and Wild Plants. *Plants* **2022**, *11*, 1081. https://doi.org/10.3390/plants11081081

Academic Editors: Oscar Vicente and Mikihisa Umehara

Received: 12 January 2022
Accepted: 13 April 2022
Published: 15 April 2022

1. Introduction

Artemisia caerulescens L. is a perennial aromatic shrub that belongs to the genus *Artemisia* L. and is included in the Asteraceae family [1].

The genus *Artemisia* L. comprises more than 500 perennial herb species, and it is widespread in the Northern hemisphere, mainly in the arid and semiarid areas of Asia, Europe, and North America [2,3]. Most of the species belonging to the *Artemisia* genus are medicinal or aromatic herbs or shrubs, characterized by a pungent smell and a bitter taste due to the presence of terpenoids, typical components of their essential oils (EOs) [4]. This genus has achieved increasing phytochemical attention for its wide range of biological activities, attributable to the presence of several classes of active compounds such as terpenes found in essential oil, phenolic compounds, flavonoids, and sesquiterpene lactones [3,5]. The *Artemisia* species have been employed since ancient times for the treatment of different disorders, including malaria, hepatitis, cancer, inflammation as well as antibiotics [2], antiparasitic, insecticidal, anti-asthmatic, and antiepileptic remedies, among others [5]. Essential oils (EOs), besides the non-volatile compounds, are important secondary metabolites obtained from several plants belonging to this genus [4]. They are complex mixtures of volatile compounds widely used in traditional and modern medicine; likewise, they are used for cosmetic and pharmaceutical applications [2]. In recent times, EOs have gained increasing interest due to consumers' preference for natural products [6] and have been used in many fields of applications, from human health to agriculture, thanks to their broad spectrum of bioactivities, which is attributable to their chemical composition [7–10].

The genus *Artemisia* comprises a wide range of species that have been subjected to phytochemical attention for their chemical diversity and EO production. These differences are attributable to their ability to colonize areas with different kinds of vegetation and ecological conditions, thus leading to different morphological and biological features [11,12]. Among the countless species of *Artemisia*, *A. caerulescens* is a wild aromatic species of great environmental interest due to its ability to grow and reproduce in selected areas of the central and western Mediterranean, which is characterized by highly saline soils [13]. This is a perennial, bushy, more or less tomentose plant, with a strong aromatic odour, very branched erect stems, and bearing both whole and 1–2-pinnate leaves. The flower heads, all with tubular flowers, are not very showy and arranged in a dense panicle (Figure 1a,b). The fruits are oval achenes, smooth and without pappus. The species is in fact considered a true halophyte [14] and is listed as a characterizing and/or diagnostic species of saline EU Habitats [15,16]. Despite species showed a remarkable interest in the environmental and conservation fields, even with evidence of some particular aspects in its behaviour such as the interaction between propagation by seed and seasonal fluctuations of salinity [17], to date, there were very few studies in literature. This information deficit affects various fields of study, including those related to its phytochemical profile and medicinal properties, which are the subject of the present research.

Figure 1. (**a**) *Artemisia caerulescens* in its natural habitat in the San Rossore estate (photo by Laura Pistelli); (**b**) ripe inflorescence (**c**) 4-week-old shoots in MSO medium; (**d**) 4-week-old shoots in 4 µM MS-BA medium.

As a part of an ongoing project, which deals with the cultivation and analysis of halophyte species growing in the salt marshes on the coast of Tuscany, the present study aims to investigate preliminary evidence from in vitro cultures of *A. caerulescens* collected in the San Rossore Estate (Pisa, Italy) in order to verify the possibility of its germplasm conservation, thus avoiding environmental depletion and allowing for the selection of the most promising cultivation lines. Moreover, biochemical and phytochemical analyses were performed on both the in vitro plant culture and different parts of the wild plant, particularly the leaves, young flowering tops as well as ripe ones.

2. Results and Discussion

2.1. In Vitro Shoot Proliferation

A. caerulescens grows spontaneously in the San Rossore Estate, in salty soil close to the sea. Few studies are available on its ecological adaptability or on its growth strategy [8,12]. Aerial parts were harvested in June and September 2021, at the vegetative and flowering stages, and then used for in vitro shoot proliferation to avoid the depletion of the wild plants for future analysis. The same "mother" plants were also used for the biochemical and phytochemical analyses.

The sterilization step was repeated for each single mother plant; the obtained results are summarized in Table 1. The initial contamination ranged between 10% and 90% of the explants and appeared a few weeks after the sterilization process. The high contamination could be due to the natural and protected habitat of the spontaneous plants and the lack of a pre-treatment with pesticides before the in vitro cultivation, as reported in some literature [18]. The shoots proliferated in either Murashige Skoog basal medium [19] (MS0) or in MS with the addition of cytokinin (MS-BA). Shoot induction was performed with the addition of cytokinin 6-benzyl-aminopurine (BA) as a unique growth regulator, which was a pivotal experiment investigating the ability of A. *caerulescens* in shoot proliferation.

Table 1. In vitro culture of *A. caerulescens*. Single mother plants (M) were used as initial explants, and their contamination was observed (%). Effect of different culture media on shoot development (%). Proliferation and length of shoots (cm) after 4 weeks of cultivation. Data are presented as means ± SD (n = 10).

Mother Plant	Number of Explants	Contaminated Explants (%)	Developed Shoots (%)	Number of Shoots/Explant	Shoots Length (cm)
M1 (MS0)	18	32	88.9	1	2.67 ± 1.7
M2 (MS0)	10	0	100	1	3.5 ±1.6
M3 (MS0-BA)	20	90	0	0	0
M4(MS-BA 2 µM)	10	10	3	3	1.2 ± 0.4
M5(MS-BA 2 µM)	10	10	5	5	1.5 ± 0.5
MS0	10	0	100	2.1 ± 1.28 ^{C}	3.18 ± 0.26 ^{A}
MS-BA 1 µM	10	0	100	8.1 ± 1.63 ^{A}	1.12 ± 0.14 ^{C}
MS-BA 2 µM	10	0	100	5.4 ± 1.08 ^{B}	1.5 ± 0.12 ^{B}
MS-BA 4 µM	10	0	100	4.6 ± 1.27 ^{B}	0.76 ± 0.11 ^{D}

The superscript, uppercase letters (A–D) indicate statistically significant differences between the samples. The statistical significance of the relative abundances was determined by the Tukey's post hoc test, with $p \leq 0.05$.

After 4 weeks of cultivation, the MS0 explants showed a lower proliferation (one shoot/explant), while the explants grown in the presence of BA showed better proliferation, with 3–5 shoots in independent experiments. However, the addition of BA produced shorter shoots (1.2–1.5 cm in length) than those of MS0 (2.67–3.5 cm) (Figure 1c,d, respectively).

Therefore, other shoots were cultured with MS, combined with different concentrations of BA (1, 2, or 4 µM). The highest proliferation of shoots was obtained with the lowest BA concentration (8.1 shoots per explant), followed by the concentration of 2 µM and 4 µM BA. The highest shoot length was achieved with the use of the MS0 medium (range, 2.67–3.5 cm), and the shortest shoots (0.76 cm) with the 4 µM BA. The effect of the addition of cytokinin is well-known to increase the number of shoots in other species. In *A. vulgaris*, the shoots proliferated in liquid MS with 0.44–8.88 µM BA, and the best effect on shoot multiplication was obtained with the addition of 4.44 µM BA, producing 85.5 shoots/explant at 500 mL flask capacity and with an average length of 12.2 cm [20]. The extraordinary data differed from our results due to the methodology and the species used. Usually, the solid medium is used for shoot induction and proliferation. In *A. granatensis,* the highest axillary shoot proliferation rate was achieved with a solid medium and the addition of 0.44 μM BA, although the shoots exhibited short lengths (0.5 cm). On the contrary, the longest length of the main shoots (2.1 cm) was achieved with MSA0 [21], which did not differ from our results on shoot induction (MS0, 2.67–3.5 cm).

However, a wide range of literature on other *Artemisia* species has highlighted the use of both cytokinins (BA) and other growth regulators in producing high shoot proliferation. In *A. annua*, BA (0.5–1.0 mg/L, e.g., 2–4 μM) stimulated the highest shoot proliferation, although other types of hormones (e.g., Gibberellin acid, GA_3) also induced the development of shoots from the tip explants [22]. *A. absinthium* nodal explants cultivated in the presence of 0.5 mg/L BA with 0.25 mg/L Kinetin (kn) produced shoots with a length (6.0 ± 0.52 cm) of 3.25 ± 0.42 cm [23], although the unique addition of BA produced the highest shoot number (4.5) when the concentration was 0.5 mg/L (e.g., 2 μM) [23]. In other reports, the proliferation of shoots has been correlated with the production of secondary metabolites. In *A. alba*, the simultaneous presence of indole-3-butyric acid (IBA) and BA in the medium promoted the development of shoots and roots and phenolic compounds [24]. Low BA concentration in combination with different IBA concentrations increased the amount of phenolic compounds as compared with the basal medium without plant growth regulators [24,25].

The different responses of the *Artemisia* species to the various culture media indicate that the pivotal results obtained in *A. caerulescens* with the unique addition of BA can be improved by the further addition of growth regulators.

No data about the rooting are reported in this manuscript since the pivotal experiments were performed to obtain higher shoot proliferation and thus to investigate the production of metabolites by in vitro shoots. Rooting experiments will be performed in future experiments and will be explored combination with various growth regulators to improve the propagation technique.

2.2. Biochemical Analysis

The *A. caerulescens* wild plants were collected during the summer period to obtain different aerial organs (leaves, and young and ripe inflorescences). The photosynthetic pigments Chlorophyll a and b (chl a 930.03 and chl b 182.27 μg g^{-1} FW, respectively; Table 2) were revealed in amounts that have already been observed for other *Artemisia* sp. such as *A. pauciflora*, *A. lerchiana*, and *A. santonica*, although they were often referred to in dry weight [12,26]. In fact, an estimation of the total chlorophyll content of *A. caerulescens* based on dry material (4.69 mg g^{-1} dry weight) is similar to that of *A. santonica* (4.8 mg g^{-1} dry weight) [12]. Inflorescences contained different pigments depending on the flowering stage: the blossoms showed higher chlorophyll content than the ripe ones (376.99 and 231.31 μg g^{-1} FW, respectively). The photosynthetic pigments were evaluated in other *Artemisia* sp. tolerant to salinity; the *A. absinthium* young plants grown in salty conditions showed similar values to those of *A. caerulescens* [27].

Other pigments, carotenoids, and anthocyanins were detected and related to antioxidant activity. Leaves of wild plants showed the highest amount of carotenoids (294.09 μg g^{-1} FW), probably associated with the photoprotection activity during the summer period and with tolerance to the salty soil area [28]. The carotenoid content in all the other examined organs was comparable as their values did not exhibit a significant difference. Concerning the anthocyanins, both stages of flowers (young blossom and ripe flowers) showed the highest amount of these secondary metabolites, which are related to the pigmentation of the flowers (Figure 1b), while they were not revealed in the leaves. Interesting results were obtained regarding the polyphenol content of the various examined organs. Both inflorescences showed the highest amount, found within the range of 17.15–18.42 mg GAE g^{-1} FW, while slightly lower content was observed in fresh wild leaves (4.62 mg). The obtained data on the polyphenolic content of wild plants were in agreement with those reported by Lee et al. [29].

Table 2. Determination of metabolites in the different *A. caerulescens* organs grown in a wild area or in an in vitro culture. Data are presented as means ± SE (n = 5). Different letters indicate statistically significant differences determined by Tukey's b post hoc test ($p < 0.05$). Abbreviations: nd.—not determined; GAE—gallic acid equivalents; CE—catechin equivalents; ME—malvin-chloride equivalents.

	Wild Plants			In Vitro Shoots	
	Leaves	**Young Inflorescences (Blossom)**	**Ripe Inflorescences**	**Shoots (MS0)**	**Shoots (MS-BA 2 μM)**
Chlorophyll a (Chl a, μg g^{-1} FW)	930.03 ± 12.19 ^{A}	376.99 ± 0.23 ^{BC}	231.31 ± 4.07 ^{C}	208.9 ± 2.76 ^{C}	448.25 ± 3.76 ^{B}
Chlorophyll b (Chl b, μg g^{-1} FW)	182.27 ± 8.36 ^{A}	114.78 ± 0.54 ^{B}	68.71 ± 0.78 ^{B}	66.6. ± 0.68 ^{B}	132.79 ± 3.76 ^{AB}
Total Chlorophyll (Tchl, 284.9 μg g^{-1} FW)	1112.3 ± 20.15 ^{A}	491.76 ± 0.30 ^{BC}	300.02 ± 4.51 ^{C}	275.51 ± 3.44 ^{C}	581.04 ± 14.54 ^{B}
Total carotenoids (Tcar, μg g^{-1} FW)	294.9 ± 2.01 ^{A}	134.39 ± 0.37 ^{B}	91.47 ± 1.61 ^{B}	60.2 ± 0.78 ^{B}	124.11 ± 3.10 ^{B}
Total Anthocyanins (TA, mg ME g^{-1} FW)	nd.	7.95 ± 0.51 ^{AB}	20.47 ± 0.89 ^{A}	1.03 ± 0.05 ^{B}	0.57 ± 0.05 ^{B}
Total Polyphenols (TP, mg GAE g^{-1} FW)	4.62 ± 0.13 ^{B}	17.15 ± 0.12 ^{A}	18.42 ± 0.24 ^{A}	0.71 ± 0.01 ^{B}	0.63 ± 0.05 ^{B}
Radical scavenging DPPH-assay (μmol TEAC g^{-1} FW)	12.25 ± 0.55 ^{B}	82.15 ± 4.80 ^{A}	71.16 ± 2.94 ^{A}	2.53 ± 0.10 ^{B}	1.59 ± 0.16 ^{B}
Antioxidant activity FRAP assay (mmol Fe^{2+} g^{-1} FW)	24.35 ± 0.86 ^{B}	120.34 ± 3.64 ^{A}	102.3 ± 1.53 ^{AB}	4.91 ± 0.06 ^{B}	4.30 ± 0.3 ^{B}

The superscript, uppercase letters (A, B, C) indicate statistically significant differences between the samples. The statistical significance of the relative abundances was determined by the Tukey's post hoc test, with $p < 0.05$.

The antioxidant activity, detected with two different assays (FRAP or DPPH), was therefore the highest in the flowers of *A. caerulescens*, followed by the leaves. The activity could be linked to the polyphenols, including the anthocyanins, since they work as scavengers of free radicals and as natural metal chelators [30], and to the carotenoids, which have achieved increasing interest in the last decades for their antioxidant properties [31]. On the other hand, the contribution of other antioxidant molecules can be included and further investigated.

Biochemical analyses of in vitro shoots were performed and referred to the different culture media. The chlorophyll content (Chl a, Chl b, and total chlorophyll) of shoots was higher in the MS-BA medium (581.04 μg g^{-1} FW) than in the basal MS medium (275.51 μg g^{-1} FW), and this effect was linked to the known influence of BA in chlorophyll production. Regarding the other pigments, carotenoids were higher in the MS-BA than in the MS medium, following the trend of the other photosynthetic pigments. Anthocyanins, belonging to the class of phenolic compounds, were more concentrated in the MS0 medium (1.03 mg ME g^{-1} FW) than in MS-BA (0.57 mg ME g^{-1} FW). In vitro shoots showed a reasonable polyphenol content (0.63–0.71 mg GAE g^{-1} FW), and the antioxidant activities (DPPH and FRAP assay) were higher in MS0 than in MS-BA (Table 2). Altogether, the contribution of cytokinin BA seemed to reduce the production of antioxidant compounds. A similar effect has already been observed in the shoot culture of *Scutellaria alpina* [32] and in the shoot culture of *Artemisia alba* [24], for which the high concentration of BA limited the amount of polyphenol compounds. The initial results obtained on *A. caerulescens* provide a basis for future investigations dealing with the optimization of the secondary metabolite production with the addition of plant growth regulators [24], or the addition of elicitors, or treatment with light, or precursor feedings, as already reported for the genus *Artemisia* [33].

2.3. *Phytochemical Investigation*

2.3.1. Headspace Analysis

The complete composition of the headspaces (HSs) emitted by both the different parts of the wild plant (leaves, young inflorescences, and ripe inflorescences) and the in vitro shoots of *A. caerulescens* are reported in Table 3. A total of 35 compounds were identified, representing 97.3–100% of the whole volatile organic compounds (VOCs).

All the volatile emissions were characterized by the predominance of oxygenated monoterpenes, which resulted significantly higher in the HS of the in vitro culture (91.6%), followed by the ripe flowers (89.1%) > leaves (87.0%) > young inflorescences (70.2%). Among this class of compounds, thujones were the main constituent; in particular, α-thujone was revealed in remarkable percentages in all the samples, ranging from 45.4% in the young inflorescences and 83.2% in the in vitro shoots, whose HS was characterized by only ten compounds. β-thujone was detected in good percentages as well: it was higher in the samples which presented lower percentages of the α-isomer. The HS of young inflorescences exhibited the highest amount of this compound (11.2%), while the in vitro shoots had the lowest (7.6%).

Monoterpene hydrocarbons were well-represented in the wild plant samples, whilst they were revealed in very low percentages (0.7%) in the volatilome of the in vitro culture. The young inflorescence HS presented the highest amount of this chemical class (27.2%), followed by the HSs of both the ripe inflorescences (9.8%) and leaves (9.3%). Sabinene was the main monoterpene hydrocarbon component detected in these samples as it accounted for 5.2% of content in the leaves and 16.1% and 6.5% in the young and ripe flowering tops, respectively.

As well-evidenced in Table 3, the in vitro culture showed a less complex chemical composition of the volatile emission as compared to that of the different parts of the mother plant, probably due to a higher production of the oxygenated monoterpenes α- and β-thujones, representing the 83.2% and 7.6% of the whole aroma composition, respectively. The higher percentage of thujones in the in vitro shoots was probably attributable to a metabolism strongly directed to their production, as Dudareva et al. [34] reported a strong genetic regulation of the biosynthetic pathways involved in the volatile formation.

To the best of our knowledge, this is the first study on the chemical composition of the spontaneous volatile emission of *A. caerulescens,* although the compositions of the HSs of other *Artemisia* species have already been reported [35]. The main VOCs identified in the analysed samples were aligned with *Artemisia umbelliformis* subsp. *eriantha,* whose main compound was also α-thujone, although it is usually present in lower amounts than the β-isomer [36]. Conversely, the volatile profile of the analysed samples was very different from that of the *Artemisia argyi* H. Lév. studied by Li et al., who reported germacrene D (28.73%), α-pinene (16.44%), limonene (12.22%), cylcofenchene (6.46%), and α-phellandrene (4.06%) as the main components [37], and differing from that of *Artemisia spicigera* C. Koch and *Artemisia scoparia* Waldst. et Kit., as reported by Demirci et al., characterized mainly by camphor (37.5%) and β-pinene (20.8%), respectively [38]. Moreover, camphor, was also reported as a major component of *A. campestris* L. (31.78%), which was also rich in 1,8-cineole (23.11%) and α-thujone (16.82%) [39].

The data of the volatile composition of the plant headspaces (HSs) was subjected to multivariate statistical analysis with the use of the Hierarchical Cluster Analysis (HCA) and the Principal Component Analysis (PCA) methods.

The dendrogram of the HCA, reported in Figure 2, showed two macro-clusters: the pink and green ones. The HS of the young inflorescences was clustered by itself in the green cluster, while the other samples were grouped together in the pink group. However, the HSs of ripe inflorescences and in vitro shoots were closer to each other than to the HS of the leaves.

Table 3. Complete chemical composition of the HSs emitted both by different parts of the wild plant (leaves, young inflorescences, and ripe inflorescences) and the 4-week-old in vitro shoots of *A. caerulescens* (n = 3. ± SD).

Compounds	l.r.i. [1]	l.r.i. [2]	Class	Relative Abundance ± Standar Deviation (%)			
				Leaves	Young Inflorescences (Blossom)	Ripe Inflorescences	In Vitro Shoots
ethyl 2-methylbutyrate	850	850	nt	- [3]	0.1 ± 0.00	-	-
(*E*)-salvene	867	867	nt	-	-	0.4 ± 0.15	-
ethyl isovalerate	859	858	nt	-	1.9 ± 0.22	-	-
tricyclene	922	927	mh	-	0.3 ± 0.00	-	
α-thujene	926	1102	mh	0.1 ± 0.07	0.6 ± 0.11	0.2 ± 0.06	-
α-pinene	933	939	mh	0.9 ± 0.18	3.2 ± 0.05	0.3 ± 0.09	-
camphene	948	954	mh	0.6 ± 0.09	0.8 ± 0.04	-	-
sabinene	973	975	mh	5.2 ± 0.82	16.1 ± 1.06	6.5 ± 2.04	-
β-pinene	977	979	mh	0.1 ± 0.02	0.6 ± 0.03	0.2 ± 0.08	-
myrcene	991	991	mh	0.2 ± 0.06	0.3 ± 0.02	-	0.7 ± 0.38
2-methylbutyl isobutyrate	1016	1017	nt	-	0.3 ± 0.03	0.2 ± 0.02	-
α-terpinene	1017	1017	mh	0.1 ± 0.02	0.2 ± 0.01	0.1 ± 0.01	-
p-cymene	1025	1025	mh	1.4 ± 0.16	0.6 ± 0.03	1.7 ± 0.25	-
limonene	1029	1029	mh	0.3 ± 0.05	0.9 ± 0.17	0.2 ± 0.05	-
1,8-cineole	1031	1031	om	1.2 ± 0.04	1.2 ± 0.06	1.2 ± 0.14	0.8 ± 0.22
γ-terpinene	1058	1060	mh	0.3 ± 0.02	0.3 ± 0.00	0.6 ± 0.10	-
α-thujone	1107	1102	om	67.9 ± 1.56	45.4 ± 0.04	78.1 ± 2.56	83.2 ± 1.41
β-thujone	1117	1114	om	8.0 ± 0.64	11.2 ± 1.80	9.7 ± 0.28	7.6 ± 0.16
chrysanthenone	1126	1128	om	2.1 ± 0.43	2.6 ± 0.54	-	-
trans-pinocarveol	1139	1139	om	0.2 ± 0.03	-	-	-
(*Z*)-tagetenone	1231	1229	om	-	0.6 ± 0.02	-	-
(*E*)-tagetenone	1240	1238	om	0.3 ± 0.02	-	-	-
*iso*piperitenone	1271	1272 *	om	0.7 ± 0.02	-	-	-
perilla aldehyde	1273	1272	om	-	0.5 ± 0.30	-	-
(*E,E*)-2,4-decadienal	1316	1317	nt	0.2 ± 0.03	-	-	-
cyclosativene	1367	1371	sh	-	-	-	1.2 ± 0.26
α-copaene	1376	1377	sh	0.4 ± 0.06	-	-	1.7 ± 0.10
β-elemene	1392	1391	sh	-	-	0.6 ± 0.11	-
2-ethylidene-6-methyl-3,5-heptadienal	1395	1395	om	6.4 ± 1.01	-	-	-
cyperene	1399	1399	sh				2.0 ± 0.68
(*E*)-β-farnesene	1458	1457	sh				1.3 ± 0.17
germacrene D	1481	1485	sh				0.8 ± 0.20
β-selinene	1486	1490	sh	0.3 ± 0.02	-	-	-
T-cadinol	1641	1640	os				0.7 ± 0.10
brevifolin	1669	1675 *	pp	0.2 ± 0.03	-	-	-
Total identified (%)				97.3 ± 0.06	100.0 ± 0.01	100.0 ± 0.01	100.0 ± 0.01
Chemical Classes				Leaves	Young Inflorescences (Blossom)	Ripe Inflorescences	In Vitro Shoots
Monoterpene hydrocarbons (mh)				9.3 ± 0.95 ^{B}	27.2 ± 0.13 ^{A}	9.8 ± 2.66 ^{B}	0.7 ± 0.38 ^{C}
Oxygenated monoterpenes (om)				87.0 ± 0.97 ^{B}	70.2 ± 0.34 ^{C}	89.1 ± 2.72 ^{AB}	91.6 ± 1.35 ^{A}
Sesquiterpene hydrocarbons (sh)				0.6 ± 0.08 ^{B}	- ^{B}	0.6 ± 0.11 ^{B}	7.0 ± 0.88 ^{A}
Oxygenated sesquiterpenes (os)				- ^{B}	- ^{B}	- ^{B}	0.7 ± 0.10 ^{A}
Phenylpropanoids (pp)				0.2 ± 0.03	-	-	-
Other non-terpene derivatives (nt)				0.2 ± 0.03 ^{C}	2.6 ± 0.20 ^{A}	0.5 ± 0.17 ^{B}	- ^{C}

[1] Linear retention index experimentally determined on an HP 5-MS capillary column; [2] Linear retention index reported in the literature by Adams 2007 [40], NIST 14 [41], and NIST Chemistry WebBook [42]; * linear retention time in PubChem [43]. [3] Not detected. For the chemical classes and for the superscript, uppercase letters (A, B, C) indicate statistically significant differences between the samples. The statistical significance of the relative abundances was determined by the Tukey's post hoc test, with $p \leq 0.05$.

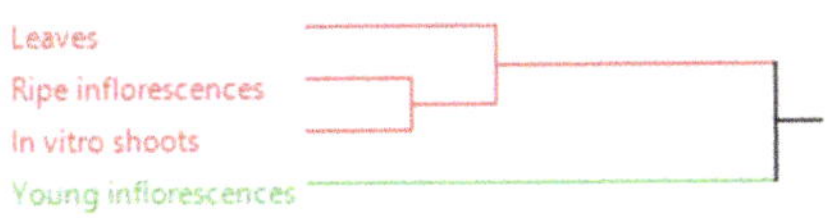

Figure 2. Dendrogram of the Hierarchical Cluster Analysis (HCA) performed on the complete chemical composition of the sample's headspaces.

The score and the loading plot of the PCA are reported in Figure 3a,b, respectively. The distribution of the samples was comparable to the partitioning of the HCA. The young inflorescence HS, which was clustered by itself in the dendrogram, was plotted in the bottom left quadrant in the score plot of the PCA (PC1 and PC2 < 0), probably for its considerable content of sabinene. The other three samples, which belong to the pink cluster, were plotted in the right quadrants (PC1 > 0); although the ripe inflorescences and the in vitro shoot HSs were located in the bottom quadrant (PC2 < 0), the HS of leaves was plotted on the partitioning line between the upper left and right quadrants, probably due to its higher content of 2-ethylidene-6-methyl-3,5-heptadienal.

Both graphs evidenced that the aroma composition of the in vitro shoots was not so different from the HSs of the different organs of the wild plant despite the fact that it was less complex because the thujones had the greatest impact on the plant fingerprint.

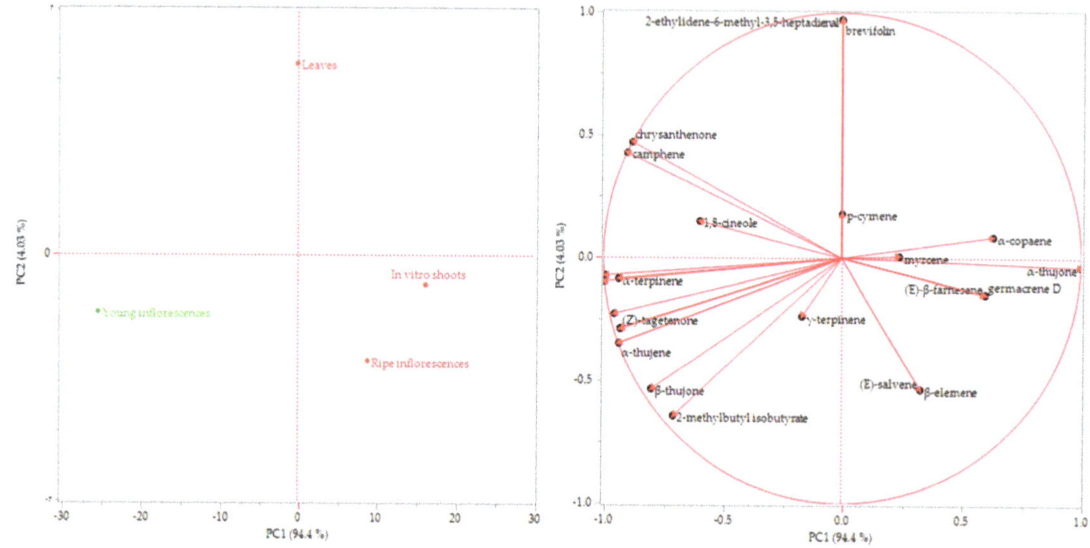

Figure 3. Score (**a**) and loading (**b**) plots of the principal component analysis (PCA) performed on the complete composition of the sample HSs.

2.3.2. Essential Oil Hydrodistillation

The EO yield was noticeable for all organs of the wild plants subjected to hydrodistillation as it ranged between 0.41% for the ripe inflorescence and 0.17% for the young inflorescences, passing through 0.35% for the leaves. This should be a positive aspect for the employment of the in vitro cultures to obtain this valuable extract. The hydrodistillation yield as well as the complete composition of the EOs obtained from the leaves, the young flowers, and the ripe ones of *A. caerulescens* are reported in Table 4. In total, 33 compounds were identified, representing 97.3–99.7% of the whole composition.

The examined samples were characterized by the predominance of oxygenated monoterpenes, proving to be significantly more abundant in the EO obtained from the leaves (74.2%), than in both the young and ripe inflorescence EOs (65.4%). Among this chemical class, α-thujone was the most relevant compound, reaching 51.5% of the whole chemical composition in the EO of the leaves, 41.7% in the EO of ripe flowers, and 39.7% in the EO of young ones. β-thujone was also well-represented in all the samples, comprising a total of 9.3–12.0%. The chemical composition of the studied EOs was in accordance with Flamini et al. [35], who reported α- and β-thujone as the main compounds and the absence of camphor in the EO of *A. caerulescens* var. *palmata*.

Table 4. Complete chemical composition and hydrodistillation yield of the essential oils obtained from fresh leaves and young and ripe inflorescence samples of *A. caerulescens* (n = 3; ±SD).

Compounds	l.r.i. [1]	l.r.i. [2]	Class	Relative Abundance (%) ± SD		
				Leaves	Young Inflorescences (Blossom)	Ripe Inflorescences
ethyl isovalerate	859	858	nt	- [3]	0.1 ± 0.03	-
α-pinene	933	939	mh	-	0.2 ± 0.03	0.1 ± 0.00
sabinene	973	975	mh	0.7 ± 0.18	1.3 ± 0.01	1.2 ± 0.02
p-cymene	1025	1025	mh	0.2 ± 0.04	-	0.3 ± 0.03
1,8-cineole	1031	1031	om	0.3 ± 0.01	0.3 ± 0.02	0.3 ± 0.01
γ-terpinene	1058	1060	mh	-	0.1 ± 0.01	0.1 ± 0.01
cis-sabinene hydrate	1066	1070	om	0.3 ± 0.04	0.3 ± 0.02	0.4 ± 0.05
trans-sabinene hydrate	1098	1098	om	0.1 ± 0.03	0.2 ± 0.01	0.5 ± 0.04
α-thujone	1107	1102	om	51.5 ± 0.03	39.7 ± 1.81	41.7 ± 0.78
filifolone	1108	1109	om	1.6 ± 0.54	-	-
β-thujone	1117	1114	om	9.3 ± 1.45	12.0 ± 0.07	9.9 ± 0.27
dehydrosabinaketone	1119	1121	om	-	0.2 ± 0.05	-
chrysanthenone	1126	1128 *	om	9.5 ± 1.24	8.1 ± 1.23	6.6 ± 0.02
trans-pinocarveol	1139	1139	om	0.5 ± 0.05	0.4 ± 0.03	1.2 ± 0.07
pinocarvone	1163	1165	om	0.3 ± 0.04	0.2 ± 0.02	1.0 ± 0.01
4-terpineol	1177	1177	om	0.4 ± 0.08	0.4 ± 0.00	0.3 ± 0.00
myrtenal	1194	1196	om	0.1 ± 0.02	0.1 ± 0.00	-
piperitone	1254	1253	om	0.2 ± 0.05	-	-
cis-chrysanthenyl acetate	1262	1265	om	-	0.4 ± 0.03	0.8 ± 0.01
*iso*piperitenone	1271	1272 *	om	0.2 ± 0.03	0.1 ± 0.02	-
trans-sabinyl acetate	1294	1291	om	-	3.1 ± 0.41	2.0 ± 0.09
β-caryophyllene	1419	1419	sh	-	-	0.3 ± 0.02
germacrene D	1481	1485	sh	0.5 ± 0.27	0.8 ± 0.15	-
β-selinene	1486	1490	sh	-	1.0 ± 0.07	1.7 ± 0.14
phenylethyl 3-methylbutyrate	1491	1490	nt	0.3 ± 0.20	-	-
α-cadinol	1654	1654	os	-	0.1 ± 0.02	-
brevifolin	1669	1675	pp	23.6 ± 1.21	27.9 ± 2.21	28.8 ± 1.07
mustakone	1687	1676	os	-	0.3 ± 0.01	-
Eudesma-4(15),7-dien-1β -ol	1688	1688	os	-	0.3 ± 0.04	-
β-nootkatol	1712	1712	os	-	0.3 ± 0.01	-
(*Z*)-lanceol	1762	1761	os	-	1.0 ± 0.14	-
methyl *iso*costate	1792	1791	os	-	0.7 ± 0.21	-
kaurene	2048	2043	dh	-	0.2 ± 0.01	-
Chemical Classes				Leaves	Young Inflorescences (Blossom)	Ripe Inflorescences
Monoterpene hydrocarbons (mh)				0.9 ± 0.22 [B]	1.7 ± 0.03 [A]	1.8 ± 0.05 [A]
Oxygenated monoterpenes (om)				74.2 ± 2.01 [A]	65.4 ± 2.74 [B]	64.7 ± 1.02 [B]
Sesquiterpene hydrocarbons (sh)				0.5 ± 0.27 [B]	1.7 ± 0.22 [A]	2.0 ± 0.16 [A]
Oxygenated sesquiterpenes (os)				-	2.7 ± 1.59	-
Phenylpropanoids (pp)				23.6 ± 1.21 [B]	27.9 ± 2.21 [A]	28.8 ± 1.07 [A]
Other non-terpene derivatives (nt)				0.3 ± 0.20	0.1 ± 0.03	-
Total identified (%)				99.5 ± 0.12	99.7 ± 0.00	97.3 ± 0.15
OE hydrodistillation yield (% *w*/*w*)				0.35 ± 0.01 [B]	0.17 ± 0.01 [C]	0.41 ± 0.01 [A]

[1] Linear retention index experimentally determined on an HP 5-MS capillary column; [2] Linear retention index reported in the literature by Adams 2007 [40], NIST 14 [41], and NIST Chemistry WebBook [42]; * linear retention time in PubChem [43]. [3] Not detected. For the chemical classes and the EO hydrodistillation yield, the superscript uppercase letters (A, B, C) indicate statistically significant differences between the samples. The statistical significance of the relative abundances was determined by the Tukey's post hoc test, with $p \leq 0.05$.

Tujones are volatile monoterpene ketones widely used as flavouring agents in the food industry. The accumulation of α- and β-thujone in the EOs is influenced by different external and internal factors, such as the plant genetic heritage, organ, and growth stage as well as weather and environmental conditions. Their biosynthesis, in fact, starts from geranyl-diphosphate (GPP) and neryl-diphosphate, through a four-step biosynthetic pathway, whose first produced monoterpene is sabinene. Despite sabinene, one of the most widespread monoterpene compound detected in the EOs, is the indirect precursor of thujones, their biosynthesis is restricted to only a few species as a consequence of plant genetic expression [44]. Unsurprisingly, since the metabolism was shifted to the production of thujones, the analysed samples showed low amounts of sabinene, ranging from 0.7 to 1.3% of the whole EO compositions.

Today, thujones are the focus of debate concerning their effect on human health. Their presence in products intended for human consumption is regulated by the European Parliament and Council and the European Medicines Agency due to reported toxic effects related to their use. However, plants containing thujones have been widely employed as natural remedies in ethnobotanical applications. Recent investigations have cleared the mechanism of neurotoxicity in these chemicals as they are modulators of the GABA-gated chloride channels, but those studies have also evidenced several potential benefits such as their immune-modulatory and anti-carcinogenic properties as well as their antimicrobial effect. The effect of these chemicals seems to be strongly dose-dependent [44].

Moreover, the analysed samples were characterized by relevant amounts of chrysantenone, accounting for 6.6%, 8.1%, and 9.5% of the total amount in the EOs obtained from ripe inflorescences, young inflorescences, and leaves, respectively. The differences in the relative content of this component were noticeable between the sample headspace and the essential oil. In fact, chrysantenone was only detected in the HSs of leaves and young inflorescences, which amounted to 2.1% and 2.6% of the total amount, respectively. Despite the fact that this molecule has not ever been identified in *A. caerulescens* before, its presence was not surprising as it has been found to be one of the major components of the EO obtained from the aerial parts of *Artemisia herba-alba,* which showed high percentages of α- and β-thujone as well [45].

Phenylpropanoids were the second relevant class of compounds in all the analysed samples, resulting in higher representation in the EOs of both the young and the ripe inflorescences (28.8% and 27.9%, respectively) than in those obtained from the leaves (23.6%). The only detected compound belonging to this class was brevifolin, which has not been detected in *A. caerulescens* before but was revealed in good percentages in the EO obtained from *Artemisia turcomanica* Gand. [46].

3. Materials and Methods

3.1. Plant Materials

The spontaneous wild *Artemisia caerulescens* L. plants were harvested between June and September 2021 in a typical salt marsh called Lame, located in the San Rossore Estate, the heart of the Migliarino San Rossore Massaciuccoli Regional Park (Pisa, Italy). According to the most recent taxonomic revisions [47], the plants chosen for the study refer to *Artemisia caerulescens* subsp. *caerulescens* (=*Seriphidium caerulescens* (L.) Soják), previously cited under the no longer accepted name of *Artemisia caerulescens* L. var. *palmata* Lam. [43,44]. Reports on the species in the collection area are catalogued and georeferenced in the Wikiplantdbase #Toscana database (2015) [48]. Five plants in June (vegetative stage) and five plants during the blooming and the full-flowering phase (September) were used for the experiments. Leaves and flowers were directly used for the analysis of the volatile emission and the extraction of the essential oil. The aerial part of the plants (nodes) was also used for the in vitro propagation. The biochemical analyses were thus performed by the collection of fresh organs (leaves, flowers, in vitro shoots) and kept at −20 °C until use. In vitro shoots and the aerial organs of spontaneous plants (leaves, flowers) were used for biochemical and phytochemical analyses.

3.2. In Vitro Shoot Proliferation

Aerial parts of spontaneous plants were divided into small nodal parts. Apical and nodal parts (1–1.5 cm) were used as explants and washed in Tween-20 for 30 min. The NaClO solution (20%) was used as a sterilization agent for 30 min, and the explants were further washed in sterilized water. These were subsequently placed in a culture medium composed of Murashige and Skoog salts and vitamins [19], 3% sucrose, and 0.8% agar, and the pH level was adjusted to 5.8 (called MS0). Other explants were placed in MS0 medium with the addition of 1, 2.4 μM 6-benzyl-aminopurine (BA, called MS-BA). The explants were maintained in a culture chamber at 23 ± 1 °C, with a 16/8 h light/dark cycle and 50 ± 5 μmol m^{-2} s^{-1} light irradiance. The subcultures were processed every 4 weeks.

3.3. Biochemical Analysis

3.3.1. Pigment, Polyphenol, and Flavonoid Extraction and Determination

Ground/powdered fresh leaves of *A. caerulescens* (0.2 g each replicate) were incubated with 10 mL of 100% methanol for 24 h at 4 °C, and the absorbance was subsequently read at 665 nm, 652 nm, and 470 nm in a SHIMADZU UV-1800 spectrophotometer (Shimadzu®, Japan). Chlorophyll a, b, total chlorophyll, and carotenoid content were determined using the proper formulas reported by Lichtenthaler [49]. Biochemical determinations of total polyphenols (TP), total flavonoids (TF), and antioxidant activity were performed on extracts obtained by the homogenization of 0.2 g of the plant materials with 2 mL of 70% aqueous methanol, then kept for 30 min in ice, and centrifuged at 14,000× *g* for 20 min. The supernatants were used for the biochemical determinations. TPC was determined using a modified protocol of the Folin Ciocalteau method [50]. The analysis was carried out on 10 μL of the supernatant, in triplicates. The incubation was performed at 40 °C for 30 min, then the absorbance was spectrophotometrically determined at 765 nm. Total phenolic content (TP) was expressed as mg of GAE per g of DW (μg gallic acid equivalents per g FW). The total flavonoid content (TF) was determined as reported by Kim et al. [51] in 50 μL of the plant sample extracts. The absorbance was read at 415 nm and 510 nm, and the concentration was expressed as mg of (+)-catechin equivalents (CE) per g of FW.

3.3.2. Antioxidant Activity

The antioxidant activity of the fresh leaves, young inflorescences, and ripe inflorescences of the wild plants and the aerial part of the in vitro shoot of *A. caerulescens* was determined by using the DPPH (2,2-diphenyl-1-picrylhydrazyl radical) and FRAP (Ferric Reducing Antioxidant Power) scavenging methods [52]. The assays were performed in triplicates.

Concerning the DPPH scavenging method, 20 μL aliquots of the methanolic extract were added to a 0.25 mM (*w*/*v*) DPPH methanol solution to reach a final volume of 1 mL. After 30 min of incubation at room temperature in the dark, the blenching of DPPH was measured at 517 nm. Trolox was used as control (2.5 mM), and the activity was expressed as μmol Trolox Eq g^{-1} FW.

The scavenging activity of the sample by FRAP assay was performed using the Szôllôsi method [53]: aliquots of 20 μL of the sample were added to 900 μL of FRAP solution. After 4 min of incubation at room temperature, the absorbance was read at 593 nm. $FeSO_4$ was used as standard, and the activity was expressed as mmol Fe^{2+} g^{-1} FW.

3.4. Phytochemical Investigation

3.4.1. HS-SPME Analysis

The spontaneous volatile emission of the fresh roots, leaves, young inflorescences, and ripe inflorescences of the adult plant and the aerial part of the in vitro plant were analysed in triplicate by HS-SPME (Headspace Solid-Phase Microextraction). The samples (2 g each) were introduced into a 50 mL glass flask, subsequently covered with an aluminium foil, and left to equilibrate for 30 min at room temperature. Then, the headspaces were analysed using a Supelco PDMS fibre (100 μm) (Supelco analytical, Bellefonte, PA, USA), preconditioned according to the manufacturer's instructions. The sampling of the headspaces were performed for 5 s for the leaves and the flowering tops of the adult plants, and for 15 s for the aerial part of the in vitro plant; the fibre was then withdrawn into the needle and immediately injected into the GC-MS apparatus.

3.4.2. Essential Oil Hydrodistillation

The fresh leaves, the young inflorescences, and the ripe ones of A. caerulescens were hydrodistilled separately with a Clevenger-type apparatus (Tecnovetro Snc, Pisa, Italy) in order to obtain the essential oil. The hydrodistillation was performed in triplicates on 80 g of leaves, 40 g of young flowering tops, and 20 g of ripe inflorescences and was protracted for 2 h. Then, the collected EOs were diluted in 5–10% HPLC-grade n-hexane and injected into a GC-MS apparatus.

3.4.3. Gas Chromatography—Mass Spectrometry Analyses

The gas chromatography–electron impact mass spectrometry (GC–EIMS) analyses were performed with an Agilent 7890B gas chromatograph (Agilent Technologies Inc., Santa Clara, CA, USA) equipped with an Agilent HP-5MS capillary column (30 m × 0.25 mm; coating thickness 0.25 μm) and an Agilent 5977B single quadrupole mass detector. The analytical conditions for both the EOs and the SPME analyses were set as follows: oven temperature ramp from 60 to 240 °C at 3 °C/min; injector temperature, 220 °C; transfer line temperature, 240 °C; carrier gas helium, 1 mL/min. For the EO analyses, the injection volume was 1 μL, with a split ratio of 1:25.

The acquisition parameters were the following: full scan; scan range: 30–300 m/z; scan time: 1.0 s. The identification of the constituents was based on a comparison of the retention times with those of pure samples, comparing their linear retention indices relative to the series of *n*-hydrocarbons. Computer matching was also used against commercial (NIST 14 and ADAMS 2007) and laboratory-developed mass spectra libraries, built up from pure substances and components of commercial essential oils of known composition and MS literature data [40,54–58].

3.5. Statistical Analysis

The analysis of variance (ANOVA) was performed using the JMP Pro 14.0.0 software package (SAS Institute, Cary, NC, USA). Concerning the phytochemical investigation, ANOVA analyses were carried out on the chemical classes of compounds for both the essential oil and the headspace analyses, and on the EO hydrodistillation yield. Concerning the biochemical analysis, ANOVA analyses were performed on chlorophyll A, chlorophyll B, total chlorophyll, total carotenoids, total polyphenols, total anthocyanins, radical scavenging assay (DPPH), and antioxidant activity (FRAP). Averages were separated by Tukey's b post hoc test. $p < 0.05$ was used to assess the significance of differences between means. Multivariate statistical analyses were also performed with the JMP software package. Hierarchical cluster analysis (HCA) was performed using Ward's method on unscaled data, with squared Euclidean distances as a measure of similarity. Principal component analysis (PCA) was performed on a 35 × 4 data covariance matrix (35 compounds × 4 samples = 140 data), selecting the two highest PCs, PC1 and PC2, obtained by the linear regression operated on the mean-centred, unscaled data, covering 94.40% and 4.03% of the variance, respectively, for a total variance of 98.03%.

4. Conclusions

In the context of climatic changes, the micropropagation technique was useful in avoiding the germplasm depletion of species found in a protected area and in maintaining biodiversity. The preliminary results on the proliferation of in vitro shoots of *A. caerulescens*, a halophyte species still not fully characterized, showed the best results with the MS medium, added with 1 μM of BA.

Biochemical analyses and antioxidant activity were investigated on both in vitro shoots and wild plants as well as the chemical composition of the spontaneous volatile emission, which, to the best of our knowledge, has never been studied before. Moreover, we also reported the chemical composition and the hydrodistillation yield of the essential oils obtained from the different organs of the plant.

Total polyphenol content and antioxidant activity were noticeable in both the inflorescences, while leaves and in vitro shoots showed lower amounts. Due to the low amount of antioxidant compounds in the in vitro shoots, further studies are necessary to improve the production of secondary metabolites.

Concerning the phytochemical investigation, the oxygenated monoterpenes α- and β-thujone were the main compounds detected in both the analysed headspaces (HSs) and the essential oils (EOs). They are the representative compounds of the *Artemisia* genus, and their biological activity is widely used in folk medicine. The EO yield was noticeable in all the organs of the wild plant subjected to hydrodistillation. The EOs were also characterized by noticeable percentages of phenylpropanoids (23.6–28.8%), with brevifolin as the unique compound, which was not detected in the spontaneous volatile emissions of the same parts of the wild plant. The high percentage of phenylpropanoids identified in the EOs and the high EO yield represent an interesting starting point for future investigation into their biological activities that will promote their employment in other possible applications.

Author Contributions: Conceptualization, L.P. (Laura Pistelli); methodology, Y.P. and L.P. (Laura Pistelli); software, Y.P.; validation, L.P. (Laura Pistelli), L.P. (Luisa Pistelli), and Y.P.; formal analysis, Y.P., M.V., and T.L.; investigation, T.L., L.P. (Laura Pistelli), and L.P. (Luisa Pistelli); resources, L.P. (Luisa Pistelli) and L.P. (Laura Pistelli); data curation Y.P. and L.P. (Laura Pistelli); writing—original draft preparation, L.P. (Laura Pistelli), L.P. (Luisa Pistelli), Y.P., and T.L.; writing—review and editing, all the authors.; project administration, L.P. (Laura Pistelli) and T.L.; funding acquisition, L.P. (Laura Pistelli) and T.L. All authors have read and agreed to the published version of the manuscript.

Funding: This research was funded by the research project HALOphytes grown in saline Water for the production of INnovative ready-to-eat salad–HALOWIN of the University of Pisa (PRA_2020_43, project number 34).

Institutional Review Board Statement: Not applicable.

Informed Consent Statement: Not applicable.

Data Availability Statement: The data used in this work are new and original, and they are fully reported in the present manuscript.

Acknowledgments: The authors thank the San Rossore Estate (Pisa, Italy) for their permission to conduct this research.

Conflicts of Interest: The authors declare no conflict of interest.

References

1. Rocha, M.I.; Gonçalves, M.J.; Cavaleiro, C.; Cruz, M.T.; Pereira, C.; Moreira, P.; Salgueiro, L.; Figueirinha, A. Chemical characterization and bioactive potential of *Artemisia campestris* L. subsp. *maritima* (DC) Arcang. essential oil and hydrodistillation residual water. *J. Ethnopharmacol.* **2021**, *276*, 114146. [CrossRef] [PubMed]
2. Nigam, M.; Atanassova, M.; Mishra, A.P.; Pezzani, R.; Devkota, H.P.; Plygun, S.; Salehi, B.; Setzer, W.N.; Sharifi-Rad, J. Bioactive Compounds and Health Benefits of *Artemisia* Species. *Nat. Prod. Commun.* **2019**, *14*, 1934578X1985035. [CrossRef]
3. Abiri, R.; Silva, A.L.M.; de Mesquita, L.S.S.; de Mesquita, J.W.C.; Atabaki, N.; de Almeida, E.B.; Shaharuddin, N.A.; Malik, S. Towards a better understanding of *Artemisia vulgaris*: Botany, phytochemistry, pharmacological and biotechnological potential. *Food Res. Int.* **2018**, *109*, 403–415. [CrossRef] [PubMed]
4. Koul, B.; Taak, P. The *Artemisia* Genus: A Review on Traditional Uses, Phytochemical Constituents, Pharmacological Properties and Germplasm Conservation. *J. Glycom. Lipidom.* **2018**, *7*, 1–7. [CrossRef]
5. Bisht, D.; Kumar, D.; Kumar, D.; Dua, K.; Chellappan, D.K. Phytochemistry and pharmacological activity of the genus *Artemisia*. *Arch. Pharm. Res.* **2021**, *44*, 439–474. [CrossRef]
6. Pieracci, Y.; Ciccarelli, D.; Giovanelli, S.; Pistelli, L.; Flamini, G.; Cervelli, C.; Mancianti, F.; Nardoni, S.; Bertelloni, F.; Ebani, V.V. Antimicrobial Activity and Composition of Five *Rosmarinus* (Now *Salvia* spp. and Varieties) Essential Oils. *Antibiotics* **2021**, *10*, 91090. [CrossRef]
7. Irshad, M.; Subhani, M.A.; Ali, S.; Hussain, A. Biological Importance of Essential Oils. In *Essential Oils-Oils of Nature*; IntechOpen: London, UK, 2020; pp. 1–14. [CrossRef]
8. Raut, J.S.; Karuppayil, S.M. A status review on the medicinal properties of essential oils. *Ind. Crops Prod.* **2014**, *62*, 250–264. [CrossRef]
9. Adorjan, B.; Buchbauer, G. Biological properties of essential oils: An updated review. *Flavour Fragr. J.* **2010**, *25*, 407–426. [CrossRef]

10. Abad, M.J.; Bedoya, L.M.; Apaza, L.; Bermejo, P. The *Artemisia* L. Genus: A Review of Bioactive Essential Oils. *Molecules* **2012**, *17*, 2542–2566. [CrossRef]
11. Ivanescu, B.; Miron, A.; Corciova, A. Sesquiterpene Lactones from *Artemisia* Genus: Biological Activities and Methods of Analysis. *J. Anal. Methods Chem.* **2015**, *2015*, 247685. [CrossRef]
12. Rozentsvet, O.A.; Nesterov, V.N.; Bogdanova, E.S.; Tabalenkova, G.N.; Zakhozhiy, I.G.; Popov, A.V. Effect of Saline Soils on the Functional State of Species of the Genus *Artemisia*. *Biol. Bull.* **2019**, *46*, 294–301. [CrossRef]
13. Lombardi, T.; Bedini, S. Seed Germination Strategies of Mediterranean Halophytes Under Saline Condition. In *Handbook of Halophytes*; 2021; pp. 1685–1703. [CrossRef]
14. Sonjak, S.; Udovič, M.; Wraber, T.; Likar, M.; Regvar, M. Diversity of halophytes and identification of arbuscular mycorrhizal fungi colonising their roots in an abandoned and sustained part of Sečovlje salterns. *Soil Biol. Biochem.* **2009**, *41*, 1847–1856. [CrossRef]
15. Biondi, E.; Burrascano, S.; Casavecchia, S.; Copiz, R.; Del Vico, E.; Galdenzi, D.; Gigante, D.; Lasen, C.; Spampinato, G.; Venanzoni, R.; et al. Diagnosis and syntaxonomic interpretation of Annex I Habitats (Dir. 92/43/EEC) in Italy at the alliance level. *Plant Sociol.* **2012**, *49*, 5–37. [CrossRef]
16. Sciandrello, S.; Tomaselli, V. Coastal salt-marshes plant communities of the *Salicornietea fruticosae* class in Apulia (Italy). *Biologia* **2014**, *69*, 53–69. [CrossRef]
17. Lombardi, T.; Bedini, S.; Bertacchi, A. Germination ecology of the aromatic halophyte *Artemisia caerulescens* L.: Influence of abiotic factors and seed after-ripening time. *Folia Geobot.* **2019**, *54*, 115–124. [CrossRef]
18. Borzabad, R.K.; Sudarshana, M.S.; Niranjan, M.H. In Vitro Plant Regeneration from Leaf Explants of *Artemisia vulgaris* L.—A Medicinal Herb. *Mod. Appl. Sci.* **2010**, *4*, 130–134. [CrossRef]
19. Murashige, T.; Skoog, F. A Revised Medium for Rapid Growth and Bio Assays with Tobacco Tissue Cultures. *Physiol. Plant.* **1962**, *15*, 473–497. [CrossRef]
20. Govindaraj, S.; Kumari, B.D.R.; Cioni, P.L.; Flamini, G. Mass propagation and essential oil analysis of *Artemisia vulgaris*. *J. Biosci. Bioeng.* **2008**, *105*, 176–183. [CrossRef]
21. Clemente, M.; Contreras, P.; Susín, J.; Pliego-Alfaro, F. Micropropagation of *Artemisia granatensis*. *HortScience* **1991**, *26*, 420. [CrossRef]
22. Zayova, E.G.; Nedev, T.A.; Petrova, D.H.; Zhiponova, M.K.; Chaneva, G.T. Efficient protocol for mass micropropagation of *Artemisia annua* L. *GSC Biol. Pharm. Sci.* **2018**, *5*, 59–68. [CrossRef]
23. Shekhawat, M.S.; Manokari, M. Efficient In Vitro Propagation by Ex Vitro Rooting Methods of *Artemisia absinthium* L., an Ethnobotanically Important Plant. *Chin. J. Biol.* **2015**, *2015*, 273405. [CrossRef]
24. Koleva, P.; Wolfram, E.; Pedrussio, S.; Raynova, Y.; Evstatieva, L.; Danova, K. In vitro culture development and polyphenolics production of *Artemisia alba* Turra. *J. Biosci. Biotechnol.* **2015**, *2*, 131–136.
25. Petrova, N.; Koleva, P.; Velikova, V.; Tsonev, T.; Andreeva, T.; Taneva, S.; Krumova, S.; Danova, K. Relations between photosynthetic performance and polyphenolics productivity of *Artemisia alba* Turra in in vitro tissue cultures. *Int. J. Bioautomation* **2018**, *22*, 73–82. [CrossRef]
26. Carvalho, I.S.; Cavaco, T.; Brodelius, M. Phenolic composition and antioxidant capacity of six *Artemisia* species. *Ind. Crops Prod.* **2011**, *33*, 382–388. [CrossRef]
27. Sharifivash, R.; Shokrpour, M. Physico-Chemical Evaluation of Some Wormwood (*Artemisia absinthium* L.) Ecotypes Under Salt Stress Condition. *J. Plant Physiol. Breed.* **2017**, *7*, 81–90.
28. Ghanem, A.-M.F.M.; Mohamed, E.; Kasem, A.M.M.A.; El-Ghamery, A.A. Differential Salt Tolerance Strategies in Three Halophytes from the Same Ecological Habitat: Augmentation of Antioxidant Enzymes and Compounds. *Plants* **2021**, *10*, 61100. [CrossRef]
29. Lee, Y.J.; Thiruvengadam, M.; Chung, I.M.; Nagella, P. Polyphenol composition and antioxidant activity from the vegetable plant *Artemisia absinthium* L. *Aust. J. Crop Sci.* **2013**, *7*, 1921–1926.
30. Pieracci, Y.; Pistelli, L.; Lari, M.; Iannone, M.; Marianelli, A.; Ascrizzi, R.; Pistelli, L.; Flamini, G. *Hibiscus rosa-sinensis* as Flavoring Agent for Alcoholic Beverages. *Appl. Sci.* **2021**, *11*, 19864. [CrossRef]
31. Rao, A.; Rao, L. Carotenoids and human health. *Pharmacol. Res.* **2007**, *55*, 207–216. [CrossRef]
32. Grzegorczyk-Karolak, I.; Kuźma, Ł.; Wysokińska, H. The effect of cytokinins on shoot proliferation, secondary metabolite production and antioxidant potential in shoot cultures of *Scutellaria alpina*. *Plant Cell. Tissue Organ. Cult.* **2015**, *122*, 699–708. [CrossRef]
33. Ali, M.; Abbasi, B.H.; Ahmad, N.; Khan, H.; Ali, G.S. Strategies to enhance biologically active-secondary metabolites in cell cultures of Artemisia–current trends. *Crit. Rev. Biotechnol.* **2017**, *37*, 833–851. [CrossRef] [PubMed]
34. Dudareva, N.; Pichersky, E.; Gershenzon, J. Biochemistry of plant volatiles. *Plant Physiol.* **2004**, *135*, 1893–1902. [CrossRef] [PubMed]
35. Flamini, G.; Cioni, P.L.; Morelli, I.; Uncini-Manganelli, R.E.; Tomei, P.E. Constituents of the Essential Oil of *Artemisia coerulescens* L. var. *palmata* Lam. *J. Essent. Oil Res.* **2001**, *13*, 125–127. [CrossRef]
36. Reale, S.; Pace, L.; D'Archivio, A.A.; De Angelis, F.; Marcozzi, G. Volatiles fingerprint of *Artemisia umbelliformis* subsp. *eriantha* by headspace-solid phase microextraction GC–MS. *Nat. Prod. Res.* **2014**, *28*, 61–66. [CrossRef]
37. Li, N.; Mao, Y.; Zhang, X. Separation and Identification of Volatile Constituents in *Artemisia argyi* Flowers by GC-MS with SPME and Steam Distillation. *J. Chromatogr. Sci.* **2008**, *46*, 401–405. [CrossRef]

38. Demirci, B.; Demirci, F.; Başer, K.H.C. Headspace-SPME and hydrodistillation of two fragrant *Artemisia* sp. *Flavour Fragr. J.* **2005**, *20*, 395–398. [CrossRef]
39. Karadağ, M.; Koyuncu, M.; Atalar, M.N.; Aras, A. SPME/GC-MS analysis of *Artemisia campestris* subsp. *glutinosa*, *Lavandula angustifolia* Mill., and *Zingiber officinale* volatiles. *Erzincan Üniversitesi Fen Bilim. Enstitüsü Derg.* **2021**, *14*, 41–49. [CrossRef]
40. Adams, R.P. *Identification of Essential oil Components by Gas Chromatography/Mass Spectroscopy*; Stream, C., Ed.; Allured Publishing Corporation: Carol Stream, IL, USA, 2007; ISBN 1932633219.
41. National Institute of Standards and Technology. *NIST NIST/EPA/NIH Mass Spectral Library*; The NIST Mass Spectrometry Data Center: Gaithersburg, MD, USA, 2014.
42. NIST Chemistry WebBook. Available online: https://webbook.nist.gov/ (accessed on 25 November 2021).
43. Pubchem. Available online: https://pubchem.ncbi.nlm.nih.gov/ (accessed on 25 November 2021).
44. Zámboriné Németh, É.; Thi Nguyen, H. Thujone, a widely debated volatile compound: What do we know about it? *Phytochem. Rev.* **2020**, *19*, 405–423. [CrossRef]
45. Belhattab, R.; Amor, L.; Barroso, J.G.; Pedro, L.G.; Cristina Figueiredo, A. Essential oil from *Artemisia herba-alba* Asso grown wild in Algeria: Variability assessment and comparison with an updated literature survey. *Arab. J. Chem.* **2014**, *7*, 243–251. [CrossRef]
46. Nikbakht, M.R.; Sharifi, S.; Emami, S.A.; Khodaie, L. Chemical composition and antiprolifrative activity of *Artemisia persica* Boiss. and *Artemisia turcomanica* Gand. essential oils. *Res. Pharm. Sci.* **2014**, *9*, 155–163.
47. Portal to the Flora of Italy. 2019. Available online: http://dryades.units.it/floritaly/index.php (accessed on 5 December 2021).
48. Peruzzi, L.; Bedini, G. Wikiplantbase #Toscana v2.1. Available online: http://bot.biologia.unipi.it/wpb/toscana/index.html (accessed on 5 December 2021).
49. Lichtenthaler, H.K. Chlorophylls and carotenoids: Pigments of photosynthetic biomembranes. *Methods Enzymol.* **1987**, *148*, 350–382.
50. Singleton, V.L.; Rossi, J.A.J. Colorimetry to total phenolics with phosphomolybdic acid reagents. *Am. J. Enol. Vinic.* **1965**, *16*, 144–158.
51. Kim, D.-O.; Jeong, S.W.; Lee, C.Y. Antioxidant capacity of phenolic phytochemicals from various cultivars of plums. *Food Chem.* **2003**, *81*, 321–326. [CrossRef]
52. Brand-Williams, W.; Cuvelier, M.E.; Berset, C. Use of a free radical method to evaluate antioxidant activity. *LWT-Food Sci. Technol.* **1995**, *28*, 25–30. [CrossRef]
53. Szollosi, R.; Szollosi Varga, I. Total antioxidant power in some species of *Labiatae* (Adaptation of FRAP method). *Acta Biol. Szeged.* **2002**, *46*, 125–127.
54. Davies, N.W. Gas chromatographic retention indices of monoterpenes and sesquiterpenes on Methyl Silicon and Carbowax 20M phases. *J. Chromatogr. A* **1990**, *503*, 1–24. [CrossRef]
55. Jennings, W.; Shibamoto, T. *Qualitative Analysis of Flavor and Fragrance Volatiles by Glass Capillary Gas Chromatography*; Academic Press: New York, NY, USA; London, UK; Sydney, Australia; Toronto, ON, Canada; San Francisco, CA, USA, 1982; Volume 26.
56. Masada, Y. *Analysis of Essential Oils by Gas Chromatography and Mass Spectrometry*; John Wiley & Sons, Inc.: New York, NY, USA, 1976; ISBN 047015019X.
57. Stenhagen, E.; Abrahamsson, S.; McLafferty, F.W. *Registry of Mass Spectral Data*; Wiley & Sons: New York, NY, USA, 1974.
58. Swigar, A.A.; Silverstein, R.M. *Monoterpenes*; Aldrich Chemical Company: Milwaukee, WI, USA, 1981.

MDPI
St. Alban-Anlage 66
4052 Basel
Switzerland
Tel. +41 61 683 77 34
Fax +41 61 302 89 18
www.mdpi.com

Plants Editorial Office
E-mail: plants@mdpi.com
www.mdpi.com/journal/plants

www.ingramcontent.com/pod-product-compliance
Lightning Source LLC
Chambersburg PA
CBHW062102220926
51850CB00010B/2860